# IN THE

# EMPIRE

# of ICE

ALSO BY GRETEL EHRLICH

Stories
*Drinking Dry Clouds*

Poems
*Arctic Heart*
*To Touch the Body*
*Geode/Rock Body*

Young Adult
*A Blizzard Year*

Memoir
*A Match to the Heart*

Fiction
*Heart Mountain*

Nonfiction
*The Future of Ice: A Journey Into Cold*
*This Cold Heaven: Seven Seasons in Greenland*
*John Muir: Nature's Visionary*
*Yellowstone: Land of Fire and Ice*
*Questions of Heaven*
*Islands, the Universe, Home*
*The Solace of Open Spaces*

IN THE

# EMPIRE
## *of* ICE

### ENCOUNTERS *in a* CHANGING LANDSCAPE

GRETEL EHRLICH

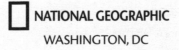

**NATIONAL GEOGRAPHIC**

WASHINGTON, DC

Published by the National Geographic Society
1145 17th Street N.W., Washington, D.C. 20036

Library of Congress Cataloging-in-Publication Data

Ehrlich, Gretel.
  In the empire of ice : encounter in a changing landscape / Gretel Ehrlich.
    p. cm.
  Includes bibliographical references and index.
  ISBN 978-1-4262-0574-3 (hardcover) -- ISBN 978-1-4262-0605-4 (e-book)
  1.  Arctic regions--Discovery and exploration. 2.  Arctic regions--Description and travel. 3.  Arctic peoples--Social life and customs. 4.  Nature--Effect of human beings on--Antarctica. 5.  Climatic changes--Arctic Regions. 6.  Global warming.  I. Title.
  G608.E57 2010
  910.911'3--dc22

                                      2009044379

The National Geographic Society is one of the world's largest nonprofit scientific and educational organizations. Founded in 1888 to "increase and diffuse geographic knowledge," the Society works to inspire people to care about the planet. It reaches more than 325 million people worldwide each month through its official journal, *National Geographic,* and other magazines; National Geographic Channel; television documentaries; music; radio; films; books; DVDs; maps; exhibitions; school publishing programs; interactive media; and merchandise. National Geographic has funded more than 9,000 scientific research, conservation and exploration projects and supports an education program combating geographic illiteracy.

For more information, please call 1-800-NGS LINE (647-5463) or write to the following address:

National Geographic Society
1145 17th Street N.W.
Washington, D.C. 20036-4688 U.S.A.

Visit us online at www.nationalgeographic.com

For information about special discounts for bulk purchases, please contact National Geographic Books Special Sales: ngspecsales@ngs.org

For rights or permissions inquiries, please contact National Geographic Books Subsidiary Rights: ngbookrights@ngs.org

*Interior design: Cameron Zotter*

Printed in the United States of America

10/WOR/1

# CONTENTS

&#x22D9;&#x22D9;&#x22D9; — &#x22D8;&#x22D8;&#x22D8;

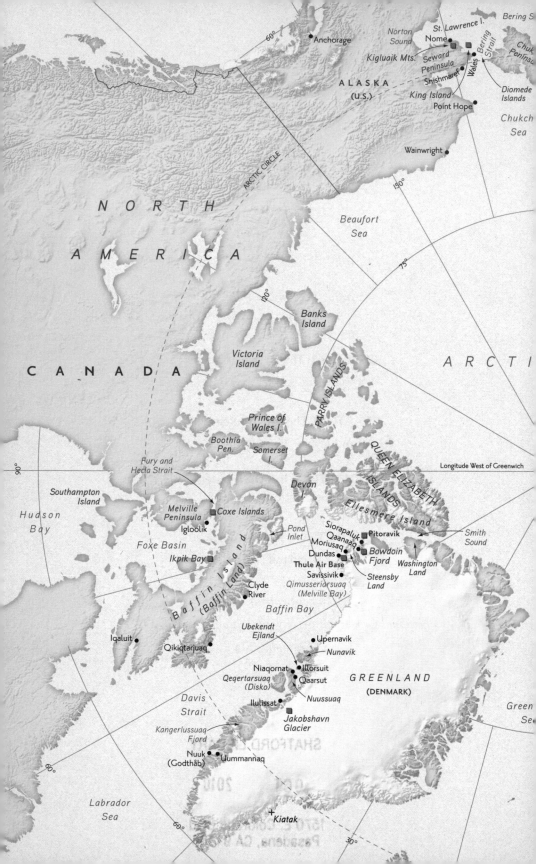

# THE EMPIRE OF ICE

CHINA

S I B E R I A

A S I A

East
Siberian
Sea

150°

New
Siberian
Islands

Laptev
Sea

120°

OCEAN

Taymyr Peninsula

RUSSIA

Pole

Longitude East of Greenwich

90°

Kara
Sea

Gyda
Peninsula

**MAP KEY**

● Selected populated place,
  settlement, or town

+ Peak

■ Point of interest

Franz
Josef
Land

Gulf of Ob

Yamal Peninsula

Svalbard

Novaya Zemlya

Barents
Sea

60°

A commonly accepted division
between Asia and Europe—here
marked by a dashed line—is
formed by the Ural Mountains
and Ural River.

75°

URAL MOUNTAINS

ARCTIC CIRCLE

E U R O P E

30°

● Indiga

● Pesha

Kola
Peninsula

White Sea

● Arkhangel'sk

Norwegian
Sea

NORWAY

FINLAND

SWEDEN

Solovetskiye
Islands

MILES
0    200    400    600

0    400    800
KILOMETERS

*Azimuthal Equidistant Projection*

# INTRODUCTION

≫——≪

ALMOST 20 YEARS AGO I stepped off a small plane onto a gravel runway and was taken two hours west by snowmobile to a seal biologist's camp on the sea ice between Cornwallis and Griffith Islands in Arctic Canada. It was spring, and the sun moved in a halo around our heads. A polar bear visited, and an arctic fox. A ferocious three-day storm battered our tent and buried our food cellar. There was no night. Days, I scratched the bellies of ringed seals the biologist caught and released. Evenings, I read from the ethnographic notes of Knud Rasmussen and watched mirages take the island archipelago and give it back again. Sun dogs shifted, clouds scuttled by as the top layer of sea ice grew soft.

The horizon was the one line that held me in place; the sea ice was evanescent. Its boundlessness, like the all-day, all-night light, was illusory and marked the coming of night. What looked smooth was rough. As the sparkle melted, my curiosity caught fire. How had people and animals thrived for thousands of years in such a place? By mid-June, the 13-foot-thick "floor" on which the camp had been set up completely dissolved, leaving no trace.

Little did I know that such a sight would become common as the world warmed, causing sea ice to decline, that by 2007 the Arctic Ocean would become a partially bald pate.

Two years later, in 1993, I began my first of 17 years' worth of journeys to northwestern Greenland, traveling by dogsled with Inuit subsistence hunters in every season. During those years I inadvertently felt the coming of a great climate shift: The sea ice on which we traveled and hunted and which was routinely 13 to 14 feet thick thinned to 7 inches. From 1998 on, I literally felt the ice go out from under my feet and the dogsleds on which I traveled.

In 2007 I received a National Geographic Expeditions Council grant to make a circumpolar journey, exploring the environment and lives of indigenous Arctic peoples and how they were being affected by climate change. Who are you? I asked them. Who were you a hundred years ago? How are you coping with vanishing ice, coastal erosion, pollution, wind and current changes, and changes in the migratory patterns of the animals on which you depend for food?

One year was not enough. A lifetime might have been more suitable. Travel in the Arctic in winter and spring is weather restricted, but I managed to get to Arctic Alaska, Nunavut, Greenland, and northwestern Russia. Sadly, I never got to my Saami friends and other hard-to-reach hunters and reindeer herders on the northeast coast of Siberia. To say that you are writing about Arctic peoples means you are also writing about whole ice-adapted Arctic ecosystems and their vanishing ice.

Arctic peoples are unique because of their environment. Isolated by ice and fierce weather, theirs represents a continuum of culture that spans tundra and ocean, ice sheets and glaciers, fjords and open-ocean ecosystems, steep coastal mountains, ice-flattened benchlands, and valleys that are verdant for the one-month-long Arctic summer.

For many seasons I was a passenger on Greenland dogsleds as the hunters roamed the frozen coast searching for seals, walruses, and polar bears. The sea ice was their highway. We lived on the ice, drinking melted multiyear ice, wearing polar bear pants, sealskin boots and mittens, and fox-fur anoraks. We ate when there was food—whatever the hunters could catch—and when they caught nothing, we humans and the sled dogs went hungry.

It was in Greenland that I saw the complexities of ice and the peregrinations of marine mammals. The men and women whose lives depended on them required a rare kind of genius, almost a second sight coupled with patience and self-discipline. Who else in the world hunts animals they can't see? How can they know where exactly under the ice these animals might be, animals that provided them with their food and clothing? Out on the frozen sea for weeks or months at a time, no day passed when they did not mention the beauty of their drifting icebergs, ice-capped rock cliffs, or the ingenuity of the polar bear.

I spent one winter in Greenland—a time when there is no light in the sky at all for four months. I imagined the top of the world as the black cap of an arctic tern. As a migrant from warmer latitudes, I tried to see this large ice-covered island from a bird's-eye perspective: the seasonal cycles of ice and animal migrations, ecosystems within ecosystems, and the circles of life and weather that bound everything and everyone together.

The first word I learned in Greenland was *Sila.* It means, simultaneously, weather, the power of nature, and consciousness. For humans and animals that have co-evolved with ice and cold, there is no perceivable boundary between a "knowing" sentient being and the strong forces of weather.

The Inuit language and its many dialects, along with the Yupik language, are spoken from northeastern Siberia to Greenland. Words and word endings in these often tell something about the environment. Place-names describe precisely the internal, deep links among earth, ice, water, animal, wind, snow, and spirit, and cautionary tales that come from that commingling. Narratives about place, people, and animals animate the unpopulated environment, and the same stories are told all the way across the Polar North. The story of the Orphan Boy told in Qaanaaq, Greenland, is the same one told in Point Hope, Alaska.

The traditional ecological knowledge of Arctic peoples is connected deeply with their icebound world. Deep ecology in the Arctic means that the sound of the walruses' clicks and whistles, humming whales, and the ululating songs of bearded seals rise through the kayak paddle. One's whole body becomes a listening post, receiving messages from under the ice.

Four months of darkness and four of bright light, storm, and stillness, and the seasonal freezing and thawing of ice, is equivalent to the circulation of human blood and thought. Animal and human minds are inextricably linked, and the ecological imagination arises from the forehead of each morning, shaped by cold and pushed into being by weather.

How we see and know a place is partly shaped by the language we speak. There is no word in English like Sila. It links natural facts and human meaning and the Inuit hunters' world of "together-doing-knowing-things" directly.

An Arctic ecosystem is small in numbers of species but highly specialized. Food webs are simple; plant and animal species are few. Their adaptation to short summers and long winters is unique. At 79° N purple saxifrage germinates, flowers, and goes to seed in three weeks, while butterflies can take fourteen years to move from chrysalis to winged creature, then live for only a few days. The narwhal's twisted tusk is really an extrasensitive tooth used to gauge weather and barometric pressure; walruses can be docile or bite a kayak in half; lemmings and musk oxen live side by side in the wide valleys at the top of Greenland and Ellesmere Island, ringed seals are sunbathers, polar bears are the top predators, little auks arrive by the thousands in swooping swarms to nest on rock cliffs almost every year on the same day. I've seen them darken the sky, their frantic fluttering like buzzing electricity, erasing all other sounds.

Pushed to the top of the world, Arctic flora and fauna have no colder latitude, no place else to go. As the climate heats up and the ice disappears, they will become extinct. Diversity is important not only to the functioning of ecosystems but also as a reminder of all that we are, as well as all that we aren't. All that "otherness" is also us.

Biological and cultural diversity is of the utmost importance. It enables land- and ice-based ecosystems to work. It ignites empathy and the power of the human imagination and functions as a survival tactic: If one language or one crop of berries fails, there are others to choose from. From polar bear to lemming, bowhead whale to ringed seal, eider duck to little auk—there's not much there, and the loss of even one element results in harsh consequences for all.

The circumpolar Arctic is topped by a white knob of ice, now melting, which is the Arctic Sea. Around it is an apron of tundra, also melting. The straits and bays between Arctic islands are the "highways" for animals and people. Though the news these days is all about the Greenland ice sheet—the remnant of the last ice age that ended about 11,500 years ago—it is sea ice where the people and animals live. For thousands of years *siku*—ice—has been a reliable lifeline for terrestrial and marine mammals, fish, birds, and humans who depended on these beings for food, shelter, watercraft, dogsleds, and clothing. The freeze-thaw cycle was predictable: nine months of sea ice (except for polynyas, areas of continuous open water) and three months of open water. Now there may be three months of ice but not necessarily in consecutive days.

Multiyear ice—ice that never melts, even in summer—has almost disappeared in the past eight years. According to scientists from the National Snow and Ice Data Center (NSIDC) at the University of Colorado in Boulder, since September 2008, 150,000 square miles of second-year ice and 73,000 square miles of multiyear ice have been moved out of the Arctic by ocean currents and winter winds. *Siquliaq*—new ice that freezes, thaws, and refreezes—is thin, salty, and unstable. Now most of the Arctic Ocean is covered by new ice.

The high Arctic is a polar desert that in the past has seen no more than two or three inches of rain a year and very little snow. Today there are flash floods and sudden, prolific snowstorms. Ice with little snow cover thickens faster; snow holds in heat. Spring and autumn storms cause wind waves that break up sea ice from beneath, so that even at minus 60°F, more ice is being lost. The ice that is "buried" in the ground and in rock—permafrost—is beginning to melt, and as it does, billions of tons of methane are being released.

Glaciers in retreat are like rivers running back up to the mountaintops where they formed. As more storms come, wind waves break up the sea ice; Greenland's ice sheet is not just melting at the top but is coming apart at the bottom. Moulins, the natural drainpipes in an ice sheet, are drilling down to the base of the ice, carrying meltwater to the glacier's sole. Basal sliding, the bottom of the glacier sliding on bedrock, causes friction, resulting in more melting. Seawater intrudes under the ice tongues of outlet glaciers, causing the tongues to break off.

THE EARTH IS A BREATHING, pulsing, lively thing, wobbling about in its own cosmic ecosystem. Oceans exhale and inhale, there are seasonal carbon dioxide sinks and sources, freezing and thawing of ice, and plants that "eat scraps of sunlight," as the science writer Oliver Morton wrote, thus bestowing on us the possibility of life. Earth is not a singular organism but a complex of living systems that maintain a planetary homeostasis—a balancing act among physical, chemical, biological, environmental, and human components. Arctic ice, including sea ice, glacier ice, permafrost, and ice sheets, drives the entire planet's climate. Weather systems are global, and the Arctic is the natural air conditioner for the entire Earth. Its seasonal blankets of snow and ice send solar radiation— heat  back into space, thus keeping our Earth temperate.

Unimaginable alterations have been occurring, the whole of which we can't see or imagine. Our carbon "sins" have pushed every major ecosystem into collapse. In March 2009, 52 climate scientists gathered by the Potsdam Institute for Climate Impact Research in Germany evaluated the five so-called tipping elements of the climate system—elements that can change "quickly and irreversibly" if the world keeps warming. Climate scientist John Schellnhuber advised all countries to commit to deeper $CO_2$ cuts of 85 to 90 percent by 2050, in order to give the world "a one in five chance of evading the worst that climate change can dish out." All the scientists agreed that major changes in the global climate system will probably occur if global warming proceeds at the current rate. These changes include the complete melting of the Greenland ice sheet, the disintegration of the west Antarctic ice sheet, a large-scale dieback of the Amazon rain forest, the increased occurrence of the El Niño phenomenon, and—if global temperatures were to rise 3.9°F—the probable collapse of the Atlantic thermohaline circulation, better known as the Gulf Stream. In addition, the snow cover over the Tibetan plateau is growing skimpy, the Indian monsoons are weakening, and the burning of grasslands and forests in Africa and South America is causing more desertification and adding to the morass of air pollution. African wildlife biologist Allan Savory said that humans have been influencing climate for a long time, long before the industrial revolution. They've been using fire as a tool to clear trees for pastureland and they've been burning grass as a natural fertilizer. "But the African savannah sequesters carbon dioxide just as forests do. Burning sterilizes the ground, and every good size fire adds as much $CO_2$ to the atmosphere as 4,000 cars," he said.

"The earth has a morbid fever," says British scientist James Lovelock, who forged the concept of Gaia in the 1970s. It is a description of life on Earth as a self-organizing, open system made up of ocean, atmosphere, land-based ecosystems, weather, geologic weathering, and the planet's surface temperatures. None of these ever operate in equilibrium—that would mean death. Instead, feedback loops, both positive (ones that amplify, called climate forcing) and negative (ones that counterbalance), maintain the interlocking physical, chemical, biological, and land-based systems. Amplifications of heat are corrected by negative feedbacks such as wind, ocean currents, ice dams breaking into the ocean, volcanoes, global dimming, or the Gulf Stream slowing down. All these cool the climate, but it is the Earth's orbital cycles, the tipping, tilting, wobbling of Earth on its axis, that has always paced the comings and goings of ice ages.

# INTRODUCTION

In 1975, the year James Lovelock introduced Gaia, scientists informed President Nixon that we were headed toward an ice age. No one paid much attention, and few predicted what was coming next. Few understood that as population and carbon emissions began to rise, the natural fluctuation toward ice ages was being erased. Global temperatures had long before begun their steady march upward, not the normal unruly ups and downs. The global temperature graph with which we are all living is shooting straight upward.

Throughout Earth's history there have been periods of greenhouse hot boxes and iceboxes, as well as short, stable interglacial periods when humans thrive and forget that nothing stays the same. Our small window of life on Earth is destined to be limited, if not by a coming ice age, then in the long term by the death of the sun itself. It's only lately that scientists have been able to track more than 750,000 years of climate history in ice, marine, and sediment cores. To understand the causes and consequences of abrupt climate shifts is invaluable for predictions about what's in store for us now, as global heating takes over. The 100,000-year cycle of ice ages doesn't include the anomalous ways climate can shift abruptly or the complexity of ocean-atmosphere chemistry, wind and sea ice, volcanism, and thermohaline disturbances and their effects.

For example, the end of one ice age can sometimes bring on another. Some 11,500 years ago, just as the world was getting warm again, the Laurentide ice sheet, which once covered the northern Unites States and all of Canada, burst and collapsed, sending the largest freshwater pulse in the past 100,000 years. Trillions of cubic feet of meltwater spilled into the North Atlantic, flooding parts of Europe and giving rise to the legend of Noah's ark. The sudden intrusion of fresh water interrupted the conveyor of warm water, the Gulf Stream. No warm winds carried heat to Britain and Europe, and immediately what is known as the Younger Dryas event began. Sea ice formed and another short ice age came into being, ending just as quickly a thousand years later.

Ice sheets are laid down, melt, and laid down again. The current Greenland ice sheet is more than two miles thick at the summit but has been in place only since the last glacial period 11,500 years ago. In the previous interglacial, boreal forests reached all the way north to the Arctic Sea, and 53 million years ago, when it was hotter there than today, the ancestors of rhinos and tapirs roamed the high-latitude mountains. Now the last ice sheet in the Northern Hemisphere is disappearing. It is starving to death because the rate of melting is exceeding the rate of snow accumulation. This is no natural fluctuation but climate forcing caused by human activity. In 2009 the air temperatures over the Arctic Ocean rose between 1.8° and 3.6°F. The area over the Barents Sea was even warmer—more than 7.2°F as a result of the absence of sea ice cover—while the Bering Sea was colder that year.

In 2004, while I was traveling by dogsled on bad ice down the coast of Greenland, two satellites, jointly called GRACE, were measuring the thickness of the Greenland ice sheet. The data received showed that the melt rate had doubled and was increasing exponentially. That same year two icebreakers were pulling 1,400-foot-long sediment cores from the bottom of the Arctic Sea in an attempt to grasp what the environment was like during the Eocene, a hothouse period 55 million years ago. Then the Arctic Ocean was 73°F and dinosaurs were clambering about in areas long since covered by ice. Sudden global heating in the Eocene was caused by shifting tectonic plates. Carbon dioxide sequestered in seafloor limestone was released and pumped into the atmosphere by volcanoes. Two teratons of carbon dioxide spewed out, raising the temperature of the Arctic by 15 degrees. By understanding that abrupt climate shift, we might be able to prepare for the changes we are experiencing now. Or will there not be time?

The carbon pollution we are pumping into the atmosphere now is no geologic accident, yet reluctance to change our carbon habits is puzzling, given that our own lives are at stake. In 2004 NASA scientist James Hansen, who headed the Goddard Institute for Space Studies, said that the Bush Administration censored him because it was thought that climate change news would be bad for business. But "business as usual" is what climate scientists say will be our undoing.

Positive feedback loops amplify heat. They are a vicious circle, a dog chasing its tail, an addictive rush that can be quelled only by the cool hand of negative feedbacks in a homeostatic balancing act, not too different from the autonomic controls inside our bodies. But when positive feedbacks stack up as they are doing now, the stampede toward enhanced heating and the melting of snow and ice become impossible to stop. What's known as the polar amplification effect is devastating the Arctic, and therefore we too will soon feel the killing force of unmitigated heat. The Arctic drives the climate of the whole world.

Groundwater, aquifers, and meltwater from snowpack will be at a premium. Yet in the Arctic, water is the enemy of ice: The more ice melts and the more open water there is, the more ice melts. As a result, ocean water, which is a heat sink, grows exponentially warmer. In the Arctic, areas of open water, even very small ones, give off mist called sea smoke that hangs over the open-water hole, insulating it further from freezing. Any hole in the ice destabilizes the ice around it, causing it to break up.

Pack ice retreats from northern coasts, leaving huge gaps of open water and dissolving land-fast ice, allowing algae and plankton blooms to occur out of season. The animals and fish that depend on the ice are thus deprived. Warmer water brings on parasites and disease in marine mammals. The Arctic supports the biggest seabird colonies in the world, but if food availability is affected, the birds cannot nest and fledge successfully. Ivory gulls, among other birds, are already in decline.

Oceans are becoming dead zones. As they acidify, they will be unable to sequester carbon dioxide. Permafrost is suddenly not so permanent—that collar at the top of the world is melting and exhaling methane.

The disequilibrium of glaciers has gone the way of ablation. They are giving away more than they get in the way of snow. Ice sheets have been tipped on their sides, and melting icebergs crowd fjords. Pack ice has retreated from shallow continental shelves where walruses delicately whisk the sandy bottoms for mollusks. Spring snow falls on broken ice. Ringed seals' denning sites, just under the snow and ice, are shattered. Pack ice has moved away from every Arctic coastline so that polar bears and walruses have to come ashore to rest, mate, hunt, and eat. Wind waves push pancake ice, cakes of newly formed ice broken by wind. The faster it drifts, the more it thins, and it's been moving fast. The sea ice extent is declining by 2.7 percent per decade, losing an average of 16,000 square miles of ice per year.

Some of the outlet glaciers that fringe the Arctic coasts of Ellesmere Island and Greenland are surging, slipping forward rapidly and breaking off in cascading floods of ice. Up on the ice cap massive pieces of ice, some weighing as much as 10 billion tons, are sliding and breaking off, causing "glacier quakes." There have been hundreds of them, and they are increasing in frequency.

# INTRODUCTION

Recently, there was a magnitude 6.1 quake in Qaanaaq. Under the ice sheet, the bones of the Arctic world are breaking.

Rapid anthropogenic climate change is full of paradoxes. Without ice, oceans, bare ground, roads, and cities are heat sinks, and the world grows hotter and hotter. A hot Earth is one that loses its groundwater. Rainfall becomes ineffective on parched, overgrazed, overfarmed land, which leads to more aridity. Persistent drought is overtaking whole continents, in the case of Australia and the farming valleys of California, the western U.S. states, the southern parts of France, all of Spain, and North Africa, where growing deserts are replacing once prolific grasslands.

Climate scientists say the Greenland ice sheet will continue to melt completely, causing global sea levels to rise by 23 feet. So much seawater, and yet fresh water will become scarce. Food production is already being diminished by fires and drought, but the human population continues to grow.

If the freshwater flux stops, the Gulf Stream and the British Isles and other parts of Europe will have frigid winters. When that cold air, driven by transpolar winds, hits the warm Pacific air, who knows what kinds of storms will erupt?

Tropical cyclones are responsible for 15 percent of the water vapor in the stratosphere. Airborne, this water vapor causes more intense cyclones, more precipitation, and many more storms that reach all the way up to the Arctic, causing coastal erosion and breaking up even more ice.

Rock and soil that have been frozen for thousands of years are melting and letting go of their carbon- and methane-rich ice. After a ten-year plateau, methane emissions increased in 2008 from the not-so-permanent permafrost around the top of the world, as well as from agricultural, oceanic, and industrial sites.

Ice is wild and appears to be what it is not. It is composed of crushed snowflakes, trapped oxygen, hard glint, captured toxins, and impermanence. It teaches us about melting solidities and how appearances can obscure the fragility of things, reminding us that because of our carelessness, the world can no longer carry us.

Care and carelessness. Why is the beauty of the Arctic important, the people and animals that inhabit it, the shifting shapes of icebergs, the wet, black mountain walls, and the layers of light—cerulean and silver threads that carry the eye a long way off? Why is the circumpolar north—which has no industry, agriculture, few roads or cars, nothing to burn, and few people—the most polluted part of the planet? How can beauty and pollution ride the same trail?

Spring in the Arctic is now toxic. The sun has burned away winter's black hood and shines hard on ice. Ice is an archivist. It traps a winter's worth of airborne contaminants from industrialized nations. Carried north up the east coast of Greenland, these contaminants circle the top of the world in the transpolar winds. "We're talkin' social justice issues, we're talkin' heavy metals, radioactivity, mercury, soot, and POPs [persistent organic pollutants]," an angry Inuit friend in Greenland told me. "We're carrying mercury from your coal plants, eating your endocrine disruptors, drinking your soot, imbibing radioactive fish and reindeer. And we are the last traditional, ice age hunting people in the world."

Greenlanders have the highest mercury count in their bodies in the world. Methyl mercury in humans causes retardation, psychological disorders, and renal damage. Black carbon causes snow and ice to melt faster, one of many subtle positive feedback loops at work. POPs are fat soluble and accumulate in the fatty tissues of marine mammals and fish. Cod is eaten by seals;

seals are eaten by polar bears, sled dogs, and humans; arctic foxes and ravens feed on what's left. Very quickly, everyone is contaminated. Immune systems are compromised. The chemicals in the flame retardants used in fabric and plastic coverings replace natural hormones, causing hermaphroditic polar bears, seals, and seabirds, and a lowered sperm count in humans. Carbon dioxide is just one of our problems—the accumulation of contaminants has overburdened our skies, our waters, our ground, and our bodies.

Now, the final paradox: The world's pollution is causing "global dimming," which alleviates some solar radiation and keeps things cooler, but it's a toxic shield that causes disease and disabilities. Take away all the pollution, and the Earth grows even hotter. A "fool's climate," James Lovelock calls it. Damned if you do and damned if you don't.

I DREAM OF ICE. Wavy, greasy, splintered, and rough. I dream that my Inuit friends are laying panes of ice on the tormented seas. I watch the ice spreading until it slides under the dogsled on which I've been traveling. I feel the calm it brings, its presence working as a balm, its white canopy giving life. I watch pancake ice break up, the world's floor going gray and rotting. I see the broken and the breaking.

This is a book about genocide: the abuse of indigenous peoples at the top of the world. This is a book about terricide: the abuse of the planet for progress and profit, paying no heed to the biological health of the world. It's like building a house with no footing. Eventually, collapse occurs.

There are things that can be done. First we need the discipline to make decisions about the natural world based on the biological health of the planet, instead of profit only. We need a cumulative carbon budget and serious penalties for every excess. We need climate change conferences to declare legally binding resolutions. There must be adequate funding for geoengineering strategies to reduce carbon emissions, and no-nonsense schemes to capture carbon—such as burying biochar (charcoal that holds carbon) in the oceans or growing blue-light nanocrystals—and new, nontoxic ways to scrub the smokestacks of power plants. Harvested sunlight, wind, and tidal power will all be bigger and bigger providers of energy, but short-term climate targets, such as urgent and rapid industrial and agricultural emissions cuts, the shutting down of all coal plants, must be acted on now. We can no longer hide from the truth. "We've dealt ourselves a bad hand," a scientist said. "We can't bluff the planet."

Surrender is not normally a word used to wage war against extinction. But surrender we must—that is, surrender our sovereignty over the planet. The interglacial paradise in which we've been living so comfortably is shifting to a world that will not be compatible with human life. Part of James Lovelock's Gaian concept was an entreaty to get people to see themselves as part of Earth's living systems, not the masters of them. As masters, we've done a poor job. We've ignored the larger workings of Earth, talking about it as if it was something apart from our lives. But take one breath and we breathe in weather; exhale one breath and we add $CO_2$ to the atmosphere.

The Earth cannot hold us. Its arms are too full. Despairing, I think about ancient ideas of beauty, such as the Navajo word *hozho,* which alludes to the total environment *(ho)*. It is a permeating beauty that includes harmony and happiness and all the steps we must keep taking to enhance hozho throughout life. Beauty saves us. Ralph Waldo Emerson wrote: "Beauty is the form under which the intellect prefers to study the world." Perhaps our sense of delight in the richness of the world has faded along with our discipline for the study of it. I hope not. My Greenlandic friend Jens Danielsen said: "The ice is not happy that the weather is going against it. I look and look and don't see the ice wanting to come back. Please tell me, whose weather is this? It is not mine."

# THE ICE NEVER SLEEPS

>>> — <<<

## THE BERING STRAIT, ALASKA

*"To many who are unfamiliar with the world of the Inupiat, it is a dark, unforgiving world . . . The land and the sea will show you its wrath if you cannot read what it tells you."*
—Herbert O. Anungazuk

IF GIVEN A SINGLE YEAR to make a circumpolar journey, it's necessary to visit some places in midwinter, when, under a dark sky and in frigid temperatures, not much is happening. That was the circumstance under which I visited Wales, an Arctic village of 150 people, across a 50-mile-wide strait from northeastern Siberia. Winter is "story-telling time," and I listened while the people of Wales talked about their lives.

Despite the modern conveniences of snowmobiles, telephones, computers, and an airport, the people of Wales, like villagers all the way up the Seward Peninsula, are semisubsistence hunters who live off bowhead whales, walruses, and seals. They also hunt eider ducks and geese, fish through the ice for tomcod, and go inland for caribou in the fall.

When ice age hunters and their families walked across Beringia, and later sailed the Bering Strait in their *bidarkas,* they continued their seminomadic hunting lives in what we now know as Alaska, all the way up to the north coast to present-day Barrow, Deadhorse, and Kaktovik.

Because Arctic Alaska is relatively low in latitude, their "larder" was much richer and more varied than that of Arctic Canada or Greenland. In some places, providing food for their families took less time, and as a result their ceremonial and material culture thrived. Much has been lost, though. Inuit people, indigenous to Arctic Alaska, are now a minority population here. Yet if you dig deeply enough, you find the essence of a culture is still there.

SIQIEAASRUGRUK (JANUARY)—the Month of the New Sun or the "Sun that Shines on Bearded Seals." Snow has been falling. Light comes late and goes early—19 hours of darkness—but the white ground and white sky bring radiance to the far north. From my high perch in a ten-story-high hotel in Anchorage, the only patch of darkness is Cook Inlet, where open water slaps the shore and pancake ice has rotted into gray rounds that drift out as the tide changes.

"The Earth possessed us," a woman from the village of Shishmaref once said. Ice shaped the Inuit mind and society, the ecological imagination, and the ethnographic landscape. Ice is womb, home, and hearse for every Arctic species. Food, shelter, clothing, spirits, shamans, masks, drum dances, watercraft, and dogsleds were elements that bound life together on the ice. To say that Inuit people and Arctic animals "adapted" to ice is to miss the point. They co-evolved with ice. Without it, humans, walruses, seals, polar bears, and whales will die. When I first began traveling in the Arctic, the sea ice was up to 14 feet thick between December and May. Now it is often no more than six inches thick in the coldest months, barely strong enough to hold a human or a polar bear.

Sea ice is dynamic, always changing. Bering Strait's pack ice grinds and gyrates, pulling away from shore toward Little

Diomede Island, flowing north along the coast toward Point Hope, then pushing south again, its stacked pressure ridges visible and audible from shore.

Pack ice is the platform from which walruses make shallow dives to scratch at sand for shellfish, the platform on which polar bears travel, hunt, and rest. Bearded seals and ringed seals haul out on its floes to catch spring sun. Pack ice is the staging ground for human hunters as well.

Seasonal sea ice is a villager's highway, a hunter's path in spring to the ice edge where bowhead and beluga whales, walruses, and bearded seals can be found. As the ice recedes and breaks up in what are now chaotic weather patterns as a result of warming temperatures, both the hunters and the hunted in this high Arctic ecosystem are threatened. According to one Inuit hunter, "The weather is so strange it can no longer be understood. That's how much it has changed."

I HAD MET MY GUIDE, Joseph Senungetuk, and his wife, Catherine, in Anchorage, where they live. I'd noticed them in the crowd at the local bookstore, and, feeling lonely on a book tour, I invited them to dinner. A native of Wales, Alaska, a tiny Inuit village on the Seward Peninsula, Joe jumped cultures early on and went to the San Francisco Art Institute. Since then, his artwork has been collected by museums. Like many Inuit people I've met, he's a man between. His book *Give or Take a Century* chronicles his childhood in Wales. After reading it, I invited Joe to be my guide and interpreter on a trip to Wales during my 2007 circumpolar journey.

The night before we leave Anchorage, we visit an old friend of his, Herb Anungazuk, another Wales native and now a National

Parks anthropologist who has the privilege of studying his own culture. That evening we go to his two-story house on a cul-de-sac. "I don't want office talk—I want to talk about the old days," Herb says. Slight of build and jittery, he is also in his 60s and happy to see his old friend Joe. We nibble chocolate cookies. "It used to be very cold in the wintertime in Wales," he says. "We always had 25-foot drifts. Remember how hard it was getting to school, sliding down drifts from the second-story window? Now it's windier and the storms are fiercer, with more south winds occurring in wintertime," he says, looking out the window as if from a village house that faced the frozen sea.

"We had good ice most of the time from December until the third week of June. Now, by mid-April or May the ice goes out, and we have years when there is almost none at all. Melting ice changes the salinity of the sea, and it's affecting the phytoplankton and fish, and that in turn affects the migrations of birds and bowhead whales, walrus and seals, and little auks and eider ducks."

He knows an old woman from Little Diomede Island, directly across the Bering Strait from Wales, who said, "Our land is changing like an old woman changes. Things don't work right anymore."

Herb lowers his head, then looks at me: "We were pretty much the same people as on Little Diomede. And we were all the same people as on the Russian side. Some Little Diomeders came from Big Diomede the night before the Iron Curtain closed them off from each other. For 40 years they didn't see or hear from their relatives. Two generations passed before they met each other again. That's how stupid wars are."

He fidgets, nervously twirling a cookie on the plate. Then he tells us that he'd been drafted and sent to Vietnam. "My brothers went too. They sat on the beach, but I was in the middle of things. A day doesn't go by without my thinking of it," he says.

His wife brings in fresh coffee as Herb continues: "There were 11 in our family. Our house is still standing in the village. It's not very big, but it didn't have to be, because we were outside all the time. When we were born, we were named for the people who graced the lives of others, even if they weren't in the family. One of my boys has eight names. When a person dies, you take that name and give it to one of your own and hope he lives in a way that would please them.

"Spring in Kingetkin [the name for Wales in Inupiat, an Inuit dialect] was beautiful. There was always a northwesterly wind about five to ten knots. It played well with the migration of marine mammals. There's a timeline for when the animals show up. The sea and the seasons have special laws, specific signs that are looked for, that tell you whether to go out hunting or not, and words for when the sea is moving into a new season."

He says that the walruses always came in about June 10 but that now they come earlier. Walruses, he says, usually eat mollusks found on the shallow ocean bottom, but some walruses were seal eaters. "I found a headless seal once floating on the water, all its insides sucked out by a walrus. Now they predate on seals even more because their ice for diving, resting on, and for hunting is gone. The drift ice is almost gone and the pack ice goes out beyond the continental shelf where it's too deep for them to dive for food.

"Walrus use their whiskers to rake the sandy ocean bottoms for scallops, clams, and other little shellfish. Once you try eating those clams from inside the walrus stomach, you're hooked! It's perfect food!"

Joe reminds Herb that their fathers grew up eating seagulls whenever there was a food shortage. Herb recalls that the best eating were the young ones with brown feathers. "Up the coast in Shishmaref, they make 'aged' fish," he says. "They put the fish they catch in a sealskin 'poke' in a pit lined with branches to age for a few months. They do the same with walrus meat, sew it up and leave it to rot. It's called *ussok*. It sounds bad, but once you eat it . . . ah . . . that's *real* food."

We look at drawings of walruses and seals, belugas and bow-heads on his wall. "The names we give animals at different times of the year are very specific," Herb says. "The walrus and the whale have multiple names. Like if one whale has a brother, or one was a yearling, or one is a bull, or a calf, or a mother with a calf. There is a name for each of them. What makes our culture special is that we have very articulate ways of describing the resources that are important to us. We knew the sea and the seasons and how the sea moved into new seasons. We knew the signs that told us when to go out hunting and whether we would die."

JANUARY 20. Anchorage to Nome. Nome to Wales. Pointed hills, curving valleys, and the sawtooth Kigluaik Range with oxbow rivers unwinding their white coils toward the sea. Joe and I are flying in a six-passenger Bering Air plane northwest from Nome to Joe's home village of Kingetkin, population 150. This will be his first visit in 17 years.

He looks out the tiny window nervously. Ahead is a cerulean wedge, the color of blue cheese—the almost dark sky into which we will be swallowed.

Joe recalls flying in the opposite direction in 1951 when his father decided to move the family from their subsistence hunting life in Wales to the gold rush town of Nome. "I was ten years old," Joe says. "There was no school beyond eighth grade in Wales. My father gave up his traditional hunting life, everything he knew about subsistence living and had to pass on to us, in order to give us five kids an education. Dad thought education was the future, not subsistence hunting. I'm still not sure."

Here and there threads of rotting ice are thrown between white-capped wind swells. The Seward Peninsula, stretching at an angle from just below the Arctic Circle, is shaped like the blade of an ax and forms the eastern core of the Pleistocene submerged continent of Beringia. In colder times the now immersed, 1,500-mile-wide land bridge linked Alaska to Siberia. Windswept barrier islands line the coast like linked arms. The beaches are treeless and gravelly, underlain by an apron of permafrost and shallow thaw ponds. The whole tundra-covered slab of continent faces the coast of northeastern Chukotka (Siberia) only 55 miles away.

Below us and off to the left of the airplane, Norton Sound and the Bering Sea are all open water. "It should be frozen," Joe says dolefully, remembering that his family often traveled to Teller by dogsled from Wales on the frozen sea. In some bays a cuticle of shore-fast ice is being battered loose by storm waves.

IT WAS SEPTEMBER 1951 when Joe's father called a local bush pilot to pick up the Senungetuk family and all their belongings and take them to Nome. Their tiny house in Wales brought $200. "When Dad finished paying for the bush plane, he had a thousand dollars left with which to start a completely new life for himself, his wife, and five children," Joe tells me.

In Nome Joe and his siblings joined the other "modern outcasts" from King Island who had been forced to relocate. "We were Eskimo hicks," he says. Life in Nome was difficult. The children, still wearing skins and mukluks, were discriminated against. Joe's father, Willy Senungetuk, a prominent hunter at home, took the only job available to him in Nome as a janitor at the local high school.

Later, Joe's older brother Ron, also now an artist, was sent to a residential school in Edgecumbe, where he and the others were forced to learn and speak English. "To defer to a second language is to reorder one's mind. The internal links between topography, weather, way-finding, and spirit become lost horizons," Joe tells me.

Out the plane window we can see the small villages at Port Clarence and Brevig Mission. King Island and St. Lawrence Island are to the south and lost in "sea smoke" and clouds. A tentlike white cloud shrouds a pointed mountain, a sign of strong winds. Wind has been one of the indicators of climate change in the Arctic: "It blows every which way and we can't tell where it's coming from next," an elder shouts over the engine roar into my ear.

Around another headland, new ice has taken hold: Now the ocean is white and the land is powder blue, but the next bay is all open water stubbed with whitecaps. Open water turns the sea into a heat sink. That warmer water, in turn, heats the air, and the temperature rises, in turn, allowing for more open sea in what scientists call a positive feedback loop, which functions like a vicious circle, amplifying rather than balancing the heat.

We land in a hard crosswind. Our short snowmobile ride to the multiuse community center where we'll sleep is all white. Joe says that nothing looks familiar, but I say that's because we can't see. We crouch outside the locked door of the building in a howling chaos of snow until another snowmobile roars up.

Ronnie dismounts and lets us in. She's short and squared off, fast and fit. "You're here about global warming?" she asks in a matter-of-fact voice. "We've got it. Had polar bears coming into the dump when the ice was bad. No one's caught nothing. No whales off Gambell [St. Lawrence Island] this year. They usually get one or two. And we saw some strange lookin' seals with long snouts and bluish skin and big eyes." She shakes her head in dismay.

"Senungetuk, huh? I know your brother," she says to Joe. "Welcome to Wales."

As she heads out the door, she turns back, "Oh yeah, you can walk around, but no one is doing it now because it's pretty much always a whiteout and in December there was a polar bear right behind here. And we're keeping our dogs tied up because there are rabid foxes everywhere."

The door slams, but not before a small mountain of snow has blown in. Ronnie hadn't even been born at the time Joe and his family left this village. She steps onto her snowmobile and roars away.

Joe wanders around the rooms of the "multi" in a daze. He is tall and wide jawed, with inquiring eyes and a growing gut. A painting by his brother Ron hangs on the wall. Joe's wife, Catherine, also an artist, has just been diagnosed with lung cancer and he's reverberating from the shock. Coming "home" to Wales is even more poignant now.

The name Kingetkin, once Cape Prince of Wales and now simply Wales, means "an elevated area." "But not very," Joe mumbles, since the front row of houses are not more than a few feet above sea level. The village is also the westernmost on the North American continent. Houses are spread along an arm of sand, bent at the elbow, facing the Bering Sea. Two protective bulbs of rock resembling two whales' heads enclose it. "That's how passing whales know this is a place where they are welcome," Joe tells me.

Behind the gravelly coast rise a rocky upland and the mountains that divide Wales from Nome. A small river cuts the village in half. Wales was once two separate villages. A sizable lagoon is a welcome resting place for migrating geese, birds, and eider ducks in the spring. The cemetery is on the mountainside, far above the wave-battered beach.

On a clear day Little Diomede Island is visible from the village, and Big Diomede, across the invisible boundary line with Russia, lies just beyond. They are stepping stones that lead to the eastern tip of the Chukchi Peninsula.

Geologically, Alaska is part of Asia. Beringia, the thousand-mile-wide grassland steppe that, in the last ice age, connected North America and Asia between latitudes 64° and 70° N, was the bridge by which the first "colonists" came to America, bringing plants and animals, diseases and languages, food and watercraft. These were the origin points of the Inuit people on their transpolar drift across Arctic America all the way to Greenland.

Now the watery strait that divides what is now Alaska and Chukotka is only 50 miles wide but remains a passageway for marine mammals and hunting people. What's left of the Bering Land Bridge is a submerged shallow shelf that reaches all the way up the coast and across to Siberia: perfect habitat for walruses, as long as there is ice.

Arctic culture is marked by continuity and subtle, precise differences: A single language, Inuktitut, spans more than half the Arctic world, from the north coast of Siberia to the east coast of Greenland, and with it go the same legends and the ice-driven culture. The hunters in today's Greenland still harpoon narwhal from kayaks as they did off King Island thousands of years ago. And each evening, in a tent on the ice or under a rocky cliff, an Inuit hunter in, say, Siorapaluk, Greenland, tells his grandchild the same orphan story of a mistreated boy who becomes a shaman as the hunter in Wales, Alaska. Yet there are many distinct dialects; hundreds of variations in the traditional tools for hunting, shelter, and cooking; and ceremonial differences. A shaman's drum in Greenland is a small oval covered in bearded seal intestine, while the same kind of drum in Wales, Alaska, is a large round covered with the stretched gut of a walrus.

"LONG AGO people did not live like we do today. You knew how something was connected to the center and therefore together," an elder from down the coast said. The "center" was spread wide to the edges of the world and the margins were part of the center. It was impossible to talk about hunting or animals without talking of watercraft, shamans, and spirits; the *umiaq* (kayak) and harpoon; trade fairs and whale dances; thought, weather, and sentient beings. It was also "a dark and unforgiving world," Herb Anungazuk said. "We knew the sea and the seasons and how the sea moved into new seasons." Death and life braided together on moving ice and tormented seas, on winter storms and in summer fogs through which no one could see.

The villages on the Bering Strait were not isolated outposts. Wales was a sentinel and capital of sorts, with smaller villages dotting the Saniq coast north of Wales along its wind-battered sand spits and sheltered inlets. Miffitagvik, Ikpek, Sinnazaat, Kigiqtaq, Sifuk, Qividluaq, Sinik, Ikpizaaq, and Espenberg were some of the villages whose people were joined by a common language and culture, by feasts and trade, but differentiated one from the other by subdialects, decorative designs, and songs.

Sea mammal hunting had come into being by the first millennium B.C., followed by the invention of the toggle-head harpoon, which detached from the shaft when an animal was hit and hooked into the flesh so that the animal could not pull away. By A.D. 900, bowhead-whale hunting was an art.

Trade was structured and intercontinental. Coastal people traded fish and marine mammals for caribou. The umiaq was seaworthy, allowing trips south to St. Lawrence Island, north to Point Hope, and east to East Cape, Siberia. Some traveled up the Colville River to trade with people on the north coast, where Barrow now is.

Summer fairs were eagerly anticipated. The inland Nunamiut traded with maritime people. Siberians traded with Americans. The people of Wainwright and Point Hope used dogs to pull *umiat* (the plural of umiaq) on rivers and traveled south to the Kobuk, down the Utokak, portaging over the Noatak to Kotzebue. There, they waited for the rivers to freeze, then returned by dogsled on river ice.

Siberians brought pieces of iron to trade. Because the Alaskans' ivory harpoon points and knife blades broke easily, animals got away. They could make the same implements with iron tips and blades that did not break, and the lust for iron grew.

Feuds broke out between the two groups. There were robberies, and in retaliation, women and children were sometimes stolen by the Chukchi. "We were a warring people," Joe told me. The Wales hunters wore armor—chest shields made of walrus-ivory slats tied together with sealskin sinew. Chukchi hunters wore iron-plated armor. Both groups were fierce. When anyone approached King Island, armor was donned by all, even if no war was imminent, and on leaving, the invited guests put the armor back on.

"Perhaps the antipathy started long ago when groups of Inuit people were pushed from Siberia. Maybe it was a time of bad weather and there wasn't enough to eat; or it could have been overpopulation. Maybe those resentments held on over the years," Joe says.

The north coast of Siberia had plenty of walrus and polar bear, but Wales was rich in fur-bearing animals, both kinds of bears, fish, seabirds and ducks, greens, and berries. The corresponding culture reflected that wealth. There were deer-antler mallets; whalebone snow beaters to get snow off boots; bone shovels; wooden fire drills to make a spark; fox-jaw amulets threaded on sealskin thongs; loon-skin and eagle-feather wands; sealskin finger masks and walrus-stomach drums; ivory belt fasteners with seal-human faces; bone, driftwood, and bead earrings; and labrets of jade and green jasper, to name just a few.

"We lacked for nothing. Every detail had been thought of. We made what we needed and we needed nothing more," Joe says.

Rituals were practiced to show respect to the hunted animals and to the rich life in the oceans and seas. "Our world was already a very special place, long before the newcomers began showing intense interest upon our land," Herb said.

A walrus gave its meat for food, its skin for houses, umiaq and kayak covers, thread, and ropes; its ivory was used for knives, harpoon points, and jewelry; its blubber was rendered for heat and light, which in turn fostered social ties and survival.

A whale could carry a small village for a year. Its meat and *mataaq* (skin and blubber) supplied minerals and vitamin C; its ribs were used as house frames and rafters; its skin was braided into sinew for sewing; and its baleen was cut into lines used for fishing. A single seal provided meat for one family and their sled dogs for a day.

*Ugruk*—bearded seal—was abundant off Wales. The animals slept with their heads above water in rough seas and were hunted from kayaks with spears. Their meat and hides were essential, and the almost translucent skin of their guts was used variously as material for raincoats, spray skirts to keep water out of kayaks, vessels, and windows. "The ancient hunter learned to strive for total perfection in the environment he shared with his prey," Herb Anungazuk said.

The women split walrus hides in half and soaked them for weeks to soften them enough to be stretched over driftwood boat frames. Villages were places of activity: Boatmaking and the sewing of dog harnesses and skin clothing, including anoraks, pants, mukluks, and mittens, was constant. Young people had an opportunity to learn from their elders until they, in turn, could teach those who came after. Extended families shared food. These were the common threads of the society.

Hunters wore double-legged caribou pants with the fur turned inside, and over them was a sealskin pant with the hair out. Under caribou parkas trimmed in muskrat and marten, they wore emperor-goose shirts, and over it all they could wear gut-skin rain shirts in stormy weather. Personal decoration for men consisted of labrets made of ivory and jasper inserted and locked into a hole just below the lower lip, as if to mimic the tusks of the walruses they hunted. Women were tattooed on chin and arms—sometimes long geometric designs, using soot from the seal oil lamp, extended all the way to the breast.

As soon as Aagruuk, the morning star, could be seen, the hunters went out by dogsled to hunt polar bears and seals with long-handled spears. January was called the month of new light. Four hours of daylight stretched to seven. Sod houses were cleaned: The walrus-gut skylight was lifted and the house's single room was aired out. New wicks made of moss were laid on seal-oil lamps, and in the ceremonial room, local shamans appeased the goddess of the sea and mediated between the polar bear, whale spirits, and the hunters. Part of the fun was to compose new songs, stage wrestling matches, and hold masked dances where the soul of, say, a seal showed through the mouth of a half-seal, half-human face, reminding villagers that both humans and the animals with whom they lived and depended upon had "personhood."

By March the whaling captains had begun working on their skin boats, carving harpoon shafts and cutting bearded sealskin trace lines for whips and dog harnesses. Old walrus-hide coverings were removed from the umiat, and in April new ones were stretched across the driftwood frames. Ice cellars were cleaned out, partly as a gesture of welcome to passing whales, and for practical reasons of space as well.

May was the heart of whaling season. When a whale was caught, its head was cut off and sent out to sea so the spirit could return home. The pressure ice was shifting, river ice melting, and seals of all kinds hauled out on the frozen slabs to bask in the sun. June was a month of celebration for the harvest. *Nalukataq*—the blanket toss in which people are thrown into the air from a stiff blanket made of sewn-together sealskins—and feasting went on. Meat was shared, and afterward families went off to traditional camps to enjoy the brief summer's warmth and the physical freedom from howling winds and snow. Seabird eggs were gathered and there was fishing. In the fall there was sometimes another, shorter whaling hunt, but by October the ice had come in, and darkness began to spread across the horizon. In December the skies were black nearly all day and all night, a darkened theater in which the shaman's drum could be heard.

Animals were sentient. They could talk and hear. Men could marry women who were polar bears or seals; children could be taken away by dwarfs. Dancers wore powerful masks that sometimes adhered to their faces with no strings. Wearing them was a way of creating passages beyond the conventional mind. Shamans had special words used only in the ceremonial house—the *qasig*—and with those words interceded between spirits, animals, and humans. Shamans could fly and dive under the ice; they could become the animal spirit, and call on others to join in; they could heal.

"We were a part of that universe because we lived within our realm, unchanged, and without damaging the delicate land," Joe says, walking from window to window, trying to get his bearings in the government building with linoleum floors and fluorescent lights. "When you lose your language, you can't find your way."

The plane that brought us earlier leaves, cutting into a sky gauzy with blowing snow. Afterward, the airport light strobes across an expanse of white as if registering a loss. "This is the new part of town," Joe says. "It wasn't here 17 years ago."

We sit in silence. He makes another pot of coffee. "Not like it used to be here with the wood stove roaring and everything done by kerosene lamp," Joe says. The sky is dark: Is it midnight or morning? It doesn't matter. Soon enough people will find out he's here and they'll come visiting, because no matter how long he's been gone, he's still one of them, and now, in his 60s, he's considered an elder.

We look at artifacts in a glass case taken from two middens in the middle of the village: a driftwood mask, a harpoon point, a carved image of a dog. "Life here 10,000 years ago wasn't much different than it was for my grandfather," Joe says, humming softly and waiting for his *kuupiaq* (coffee) to boil. "When white people started coming on strong up here, that's when we started to die."


EVENING. Davis Ongtowasrok comes to take us to his mother's house for dinner. She is one of three elders left in Wales. Davis is in his early 40s and still single. Toothless and gaunt, he looks like an older man, testimony to the rural health care system. When his father died, he dutifully moved in with his mother, Faye.

"Mom used to be the bilingual teacher," he tells us as we put on mittens, parkas, and beaver hats. With the wind chill it's nearly 40 below. "Her education went to the fourth grade, Dad's to the eighth. Now the only time we hear Inupiat being spoken is when the elders come around with visiting dance groups. I can only understand the basics. No one young talks our language anymore."

The ride to Faye's house is snow blasted. I sit in back of Davis, and Joe rides in the sled pulled behind. "How can you see where you're going?" I ask Davis. He gives me a toothless smile: "I can't, but the 'snow-go' knows the way." We flail through deep snowdrifts and quickly arrive at Faye's hidden entryway. Blowing snow sears my face. Impossible to see. On the way to the house, Joe and I fall, yelping, laughing, and gulping snow. We're like newborn seals with no feet and closed eyes.

At 78 Faye is bent but vigorous. She has an impish smile and scuttles around her cramped kitchen cutting up piles of reindeer meat and fermented walrus with her *ulu:* "I still get ice from the lagoon and driftwood from out on the beach with my sled," she says, dropping the curved blade to shove a few more sticks into the ancient woodstove.

"I was born in a sod house. It was only about five feet high. It was nice and quiet, no sound of wind. Warm, too. Grandma and the other grandma, Mom and Dad, a blind aunt, and all us kids, 13 of us, lived there. We used to sleep on the floor with reindeer mattresses. In the summer we'd take them to camp and beat the hides with sticks, and those bugs just fell out! In the summer I'd go picking berries, and still do. Everyplace around here is good. Yes, it's nice here."

The stove ticks in time to no music as the room warms slowly. We are silent, all sitting together at a blood-encrusted table crowded with carving tools and a wad of whale sinew for sewing.

Faye serves platefuls of boiled reindeer meat and fermented bearded seal on pieces of cardboard. Meltwater brought from a thaw pond is scooped into plastic cups. "Before we had glass nursing bottles, we used bearded-seal intestines and squeezed the milk out of one end," Faye remembers.

The reindeer is chewy. She sprinkles soy sauce on it. "When we're not eating reindeer, there's some good flounder at this time of year. I make a small igloo to keep the wind off my back and lie with my head at a lead [an open lane of water] in the ice and spear them. All the year, though, we are eating walrus, seal, and polar bear. I boil the bear claws. They're good. But if you eat the bear's liver, your hair will fall out!" she says, laughing, touching her thinning hair.

Out one window there's a view of the shoreline. Out the other is a big snowdrift that flanks the tiny house. Faye shakes her head dismissively. "In the old days there was so much more snow, drifts so big the windows were covered all winter, and snow tunnels led outside to the door.

"In April we'd clear the snow away from the windows to let the sun through," she says, eyeing the curving drift that embraces the house. "This last summer the wind was one day south and the next day north. Changing every day. And the clouds changed color too. I noticed it. They used to be real white all year."

She jumps up, stirs the fire, and brings more meat. We chew in silence. She points to a photograph of a man and a woman, Joe's parents. "I remember your father, Willy," she says. "We missed you when you pulled out of here. Your father was a good man. The rest of us didn't go to much school. Our school was ice, walrus, polar bear, ugruk, and reindeer. Reindeer were brought here by Mr. Lopp, the missionary who lived up on the hill. But to get those animals, you had to get religion too.

"When we had reindeer, we had wolves. They were bad. They ate only the tongue. Must not have been very hungry. We had so much food here."

What Faye saw of the world was what passed by on the Bering Sea. She remembers an umiaq full of "ladies" from Little Diomede Island. "They were smoking big pipes. They came ashore and made doughnuts for us fried in seal oil."

She says her father always took one teaspoon of seal oil every night before he went to bed because there were no doctors or medicines then. "But it didn't do no good, because he died when I was a teenager in the 1943 flu epidemic. He got sick on January 2nd, and the next day he was dead."

Faye takes a fading photograph from the wall. In it are 50 or 60 children lined up in three rows. "These were the orphans of Wales," she says. "That's all the people left after the flu came through."

Joe lifts his heavy spectacles to inspect the faces: "My grandparents died of the flu too, leaving my mother behind," he says as he scrutinizes the photograph. "That's her," he says, pointing to one of the thin figures. He stands back, closing his eyes for a second: "My mother was only a baby then. When a neighbor found her, she was sucking the breast milk of a dead woman."

He sits heavily on a three-legged stool and leans forward on his elbows: "There weren't many people here to begin with, and we needed everyone we had. The whites brought religion and sickness and not much else. We had good lives. We had ice and lots of food. We didn't need anything from the outside world."

Epidemics came to the people of Arctic Alaska with the explorers, whalers, traders, and missionaries who sailed by or came to stay. In the mid to late 1860s, an epidemic of "red sickness"—once thought to be measles but now known to be scarlet fever—struck Point Hope, Kotzebue, the upper Noatak, and Wales.

One group of coastal hunters and their families, the Pittagmiut, who numbered between 392 and 472, disappeared completely. They lived in 16- to 20-family groups on a ten-mile stretch of coast between the point of Cape Espenberg and the mouth of the Espenberg River. In the spring they hunted seals at the cape; summers, they fished for salmon on the Immachuk River and laid out long sealskin nets at Cape Deceit to catch beluga whales. In the fall, they forced caribou into a lake, then hunted them with spears from kayaks. Winter houses were dug 12 feet into the ground, lined with driftwood timbers, with a bearded seal-gut skylight cut into the peat moss roof. Precontact, the Pittagmiut thrived. They had lived on Kotzebue Sound for 13,000 years.

Waves of illness and famine kept coming. These were people who knew about death from starvation, accidents on the ice, childbirth, and more but had never heard of measles, smallpox, influenza, TB, or pneumonia, and they were defenseless against them. They couldn't know that the ships that began passing by in the 1800s were ships of death. Besides infectious diseases, the whalers also brought tobacco, firearms, and alcohol. There were stories of famine resulting from indolence caused by drunkenness, and famine caused by people too sick to hunt.

In the year 1900 a whaling ship anchored off Barrow, Alaska, brought influenza to the village. Two hundred people died within the first week. Two years later, more than a hundred residents in Barrow died of measles. The advent of the worldwide influenza pandemic began in 1918. Fifty million people died around the world, and of those, the indigenous populations were hardest hit.

"The big sickness" arrived in Wales via dogsled. The once-a-month mail run from Nome stopped at Teller, Wales, and Shishmaref, then returned via the same route. The day the mail arrived in Wales, more than mail was being carried. On the sled was the body of a Wales hunter who had died of the flu and was being brought home for burial. The man's relatives took the corpse from the sled and buried him. Those who came near were sick by nightfall and dead by the next evening.

"I remember stories my parents and aunt and uncle told me about that time. Being young when I heard those stories, I just never understood the full impact and horror, the death they saw. I just remember them telling about people getting sick and dying. It was hard to picture all the dead people my grandparents saw," says Winton Weyapuk. He's the town environmental specialist, a shy bachelor with a shiny forehead and aviator glasses who also writes poetry. He says he is looking for a girlfriend willing to live in Wales.

Winton's grandfather was just a boy standing outside his family's sod house when the village men came to get his parents. They had died and had frozen in place after the blubber lamp went out because his mother had grown too weak to keep the moss wick lit. The boy watched the men pull her, then his father through the *iqaliq*—the skylight. He had not eaten for days because there was no one to hunt and provide food for the family.

The boy saw piles of dead bodies with hungry dogs tearing at them. There were bodies everywhere. Some were half naked. One was twisted, with an arm half raised from the pain of sickness and fever. As more adults died, there was less and less hunting. Soon, there was almost no one left to provide food for humans and dogs. Those who did not die of influenza died of starvation. It was getting to be the cold time, and every day, more and more houses went dark. He remembered a day of huge blasts. The villagers had dynamited two gaping holes in the permafrost to be used as the final resting place for the dead.

The boy and his siblings were taken to an aunt's house. Then both the aunt and uncle died too. Now orphans, the children continued living alone in the house, begging for scraps of food from neighbors. More than 200 adults had died. There were more children than adults alive in the village of Wales.

Someone finally moved the orphaned children to the school-house, where the adults who were still alive made sure everyone was fed and had a place to sleep. But the infection spread quickly. More than a third of Wales's 600 to 700 residents died the first week.

The cause of the illness was unknown. It was thought that a bad spirit had come across from Siberia to kill them. In villages to the south, epidemics were said to come from the moon. An eclipse meant an epidemic was on the way, and shamans worked hard to appease the angered spirits that had caused the illness. The length of the eclipse foretold the severity of the epidemic.

After the flu ravaged Wales, help from the outside did not come for three months. The arriving missionaries forced what was left of Wales's adults to find partners and have Christian weddings so they could "legally" adopt the remaining orphans—this, in an Arctic culture whose social structure was based on collaborative living, where both marriage and adoption were informal arrangements, where children in need were always cared for.

Joe recalls, "My father was separated from his two brothers as a result of these forced adoptions. They were given to three different sets of foster parents. Then the missionaries who entered their names entered only their personal names and left out the family name altogether. In one stroke, my father's family disintegrated.

But that's how it's been. The Europeans regarded us as savages who didn't need names. They isolated us so they could take our land. They Christianized us to make us more pliable. They introduced a value system that regarded profit as the highest good, one in which the human-animal-nature matrix came last. They befouled us with tobacco and alcohol, and the resulting ills were used as an excuse for the missionaries to scold us for being bad."

JANUARY 20, 10 A.M., minus 2°F, wind, 50 mph. "This is a story about ethnocide," I say to Joe, then apologize for saying the obvious. Joe smiles. His eyebrows are raised, his mouth downturned, his salt-and-pepper hair sticking straight up. He cocks an ear as I continue: "I mean the dire harm humans keep doing to humans. Last peoples to First Peoples, and the dire harm we are doing to this planet."

Joe points with a sly grin to the T-shirt he's wearing. Pictured on the front is a row of Indians and the words "Terrorism. We've been fighting it since 1492." He smiles, but the bitterness goes deep. "Today our life can be described as one of existence within the restricted framework of an alien civilization. The aliens who came among us have taken and taken and still it does not end. Alaska today is like a walrus sitting on the visible part of an iceberg, grunting slogans of provincial patriotism, refusing to recognize the massive three-quarters of submerged and moving ice," Joe says. In other words, the part of the iceberg that represents traditional culture.

The midmorning darkness is pixilated by horizontal grains of white. Part of the problem is that Joe and I haven't really "seen" the village yet. The whiteout is all. We're advised to stay put. Without a snowmobile or a dogsled it's hard to move around without getting lost. From this new part of town built since Joe was last here, there appears to be no dwelling, no human, no animal, no village at all. "But the ghosts are around here, somewhere," Joe says. He makes another pot of kuupiaq in his Italian stovetop espresso pot and "nukes" instant oatmeal.

Soon word gets out that Joe Senungetuk is here and people begin dropping by to visit. The *pular*—people visiting each other—is an old village custom. Food and warmth is always offered, and an ear for stories.

Ray Seetook is one of the first to come. He wears coveralls and a baseball cap. Perhaps Joe sees in Ray what he might have become if he had stayed in Wales. A village elder at age 67, Ray is one of four whaling captains. He immediately begins talking about the odd weather. "We didn't even get snow until January. Usually snow comes in December, and the shore ice just formed last month, in the third week of December. Unusual weather. Ever since last spring the winds have changed quite a bit and our shore ice has already started to rot. It's been blowing 50 miles an hour and I've been staying inside like a squirrel for the last two days!

"Last month," he continues, "we saw small birds around. Maybe 10 or 20 of them. They were dark, grayish, smaller than snowbirds [snow buntings]. They were by the pond. Then we saw a couple of hawks. Never saw birds here in wintertime."

He tells us that in spring the seagulls come just when the bowhead whales arrive. Last year his was the only whaling boat to go out. He harpooned a whale, but it went under the ice. His sons were able to find it because the ice was so thin. "One used an auger and the other son had a *tuuk*, an iron bar, and we finally got it out. Oh, we were happy. It'd been lost for three days," he says. He recalls the biggest whale he got at 47 feet long. "I dedicated that one to Mom. It was so big, for a bowhead, it almost tipped us over."

I refill his coffee cup and bring out some dried fruit. Ray adjusts his glasses, leans closer, and speaks very slowly: "I can feel it, when a whale or a polar bear is out there. I'll be working somewhere, doing something, and suddenly I'll just feel it and I tell my crew, 'Hurry up, we have to go out now.' " He sits up straight and stirs his coffee.

"We've always had plenty of food. But so many young people these days, all they do is this." He makes a gesture of typing. "What's that word? Oh yes, typing. They go to their computers." He shakes his head, not understanding why they wouldn't rather be outdoors where they could wear sealskin pants and mukluks, carry driftwood harpoons with ivory handles, and put out seal nets. "No one has these things anymore. But I teach my sons everything I know."

He talks about the importance of remembering the covenant between humans and animals, of the necessity for awareness and respect. One day he shot a bear that wouldn't die. "It didn't even get wounded," he said. "Then it turned around and looked at me and I saw a black mark on its rump. My dad told me never to shoot a bear with such a mark. I was careless. I felt so bad. That was another kind of bear, you know, the kind that has come to live among us, the kind that can't be killed at all."

In the old days animate beings had a dual existence. In some places, like Nelson Island, animals were thought to be the descendants of an all-encompassing ancestor in whom there was no boundary between human and animal. The one was incomplete without the other. They belonged to the same circle, mouth to tail, tail to mouth. That is why one could take the appearance of the other. The passageway between them was open. The way was cleared. Dances and feasts were common then. Masked dances were offerings of prayer for times when food was scarce. The ceremonial masks, made by the shaman, were used for healing and gaining awareness. Such efforts were equated with peeling back the skin of an animal to reveal the *inua,* the soul of the fused human-animal being. Now the Earth is thick—that is, noisy, congested, and secular. There seem to be plenty of boundaries, but no trans-species passages.

When Ray leaves, Joe wanders around the rooms of the community center like a caged lion. "We're like sitting ducks," he says. The whiteout intensifies. He puts his nose against glass: "Even if we could see, there's probably not much to see. The old ways are gone."

Joe stirs water into a packet of freeze-dried soup. Later, he starts remembering. He remembers his father's skin boat, his umiaq, made of two and a half split walrus hides. The skins lasted two years, he tells me, three at the most, and the old ones were recycled and used to patch the holes.

He remembers knives made of carved ivory being thrust into a ringed seal's heart. He remembers eating fresh walrus breast: part meat, part milk. He remembers dances that lasted a week or a month, before the idea of schedules was imposed from the outside. He remembers a ghost in the family's shed down five snow steps from the house where the skins and hunting gear were kept. The ghost was a neighbor lady who sat in the corner chewing tobacco and sewing.

He remembers the Wales man who fell in love with a woman up the coast from Wales at Cape Espenberg, who drove his dogsled out onto the moving ice in the spring—he knew which way it would drift—and used it as a way to visit his girlfriend.

He remembers a person on Little Diomede Island who became a walrus. When that walrus-person returned to live among humans, he found he could no longer stand their smell, so he lived alone at the edge of the village. "Maybe I feel a little bit like him now," Joe says. "Not repulsed, but apart from everyone here."

Early stories of life along the Bering Strait shaped the mind, as much as humans shaped the story. A man named Apakak from Nunatak River said: "*Tulungersaq,* Raven, formed all life in the world. He began in the shape of a man and groped around in blindness. He was squatting in darkness when he discovered himself. Where he was he did not know, nor did he know how he had come to be there. But he breathed and had life. Darkness was all around him and he could see nothing. He felt with his hands. The world was clay, everything around him was dead clay. He passed his fingers over himself, and felt his face, nose, eyes, mouth. He was alive."

The multi is empty and the lights are out. Snow light flickers, a kind of polar cinema. Joe says Apakak's origin story is also a description of an artist at work. "We all live in darkness; we all made things here in Wales, we are all blind most of the time. Some things we made were for survival and some for sheer delight. One was not deemed better than the other."

It comes like a flood, Joe's remembering. He remembers the bowhead whale and bearded-seal hunts in the spring and fall, hunts that were central to life from Wales to Point Hope. As soon as breakup occurred in April or May, a "road" was cut though pressure ridges and the umiat were dragged out to the open leads beyond the ice edge. Bearded seal, ugruk, were especially abundant. The women processed the blubber and meat and preserved them in seal-intestine bags for the winter.

To be hunted and to hunt, to eat, share food, to thrive and be abundant, to shake the mind with mask dances and animal stories represented a continual, trans-species cycle of necessity, generosity, and gratitude. Once broken, the essential bonds of every Inuit nation fell apart. Kirk Oviok from Point Hope said: "The whales have ears and are more like people. The first batch of whales seen would show up to check which ones in the whaling crew would be more hospitable to be caught. Then the whales would come back to their pack and tell them about the situation, stating, 'We have someone available for us,' " as if to say they were looking for a hunter to take them, willing to give themselves as gifts to the people.

Finally, this extraordinary story told to Joe by his father: "When the whaling people came to Wales, they were bartering with us. The captain was kind of pompous and had a new flint-lock pistol that he was showing to the people. They had never seen a gun. He guessed we were warlike, and he was right. We didn't want outsiders to come and tell us how to live.

"The captain looked at the Inuks who were greeting him and picked out one guy, brought him up on deck, and said, 'We come from a place with lots of power. See that seagull up there?' He pointed the pistol at the bird and shot, and the seagull fell out of the sky.

"The Inuk watched but didn't say a word. Instead, he took out his knife and started sawing around his own neck. He pulled his hair up and cut off his own head. Holding it in his hand, he climbed down the rope ladder, dipped the stem of his neck into the sea, climbed back up, lifted his head onto his neck root, rubbed around on his neck with his hand, and healed the wound. Then he said, 'OK. I saw what you can do. Now you've seen what we can do,' and walked away."

WHEN THE WIND CALMS, Joe and I walk. "My name, Senun-getuk, is really spelled Sinanituq. It means 'a person who likes

THE ICE NEVER SLEEPS

to follow the shore.' My Inupiat name is *Inusunaaq*—'one who would like to live longest.' Maybe you could say that I am a long-lived shore-walker."

Across the half frozen lagoon, over the wooden bridge and the Village River that divides the settlement in half, we walk toward the Old Lady of the Mountain in the distance. Most of the buildings are new: a high school, a store, and houses built so close to the water that they are now exposed to the ravages of erosion. Joe is pacing in front of three old houses. They are boarded up and splintered by cold and wind. He's trying to remember which one was his but is having difficulty.

A man from St. Lawrence Island described his house: "Summertime we live in just the frame covered with walrus skin. Then, when cold weather come, move into what you call 'in-the-ground house.' Three sides place for sleep and one above, just something like a shelf. Then what you call an entrance, way to go in, kind of a narrow hole toward the sea, and build an extra frame of wood and cover on top, cover with sod. Leave just a little hole on the corner for come out, go in. Then we stay there until the month of, sometime January, February. Come out after that."

When Joe was born in the 1940s, his family no longer lived in a sod house but in what they called a lumber house. "There were five of us children and two adults living in a house with a single main room 10 by 14 plus two storage rooms tacked on in the front. Lumber was scavenged from the beach. There was a woodstove and a shelf or two for drying mittens and mukluks. The bunk bed where my sister, brothers, and I slept was four feet wide and five feet long. One bed accommodated all five children: three above and two below."

In Wales there were four, sometimes five winter *qasiit* (the plural of qasig)—dance houses. The entry, a long, low passageway, was paved with whale vertebrae, the main room made with whale-rib rafters and roofed with cakes of ice. The room was dark. Six or seven drummers sat on one side beside huge stacks of whale, walrus, and bearded-seal meat, and buckets full of snowballs used to cool children's faces and to revitalize dried-out drum skins. An elder tended a large seal-oil lamp. The heat was oppressive. Youngsters wore no clothes. Halfway through the six-hour-long ceremony, gifts of seal gut, carved ivory, mukluks, mittens, and trade items were given out. Puppets carved from driftwood provided shadow play. Sometimes a shaman drummed while a masked whaling crew danced, reenacting the hunt. Stickpaths, shaman's apprentices, were called on for help. A man might be speared, as if he were the whale, and blood was seen pouring from his back, but once the masks were removed, no blood was visible.

Joe remembers only a single qasig, perhaps the last one to exist here, made of driftwood, walled with sod and lined with sealskin and caribou or reindeer hides. As in all the houses, a seal-gut or walrus-gut window let in light and could be removed to let smoke out when a fire was built inside or to let in fresh air. In the summer, people came and went through that top hole.

Much of the village activity occurred in the dance room. On a winter day like this, when the snow was blowing hard and it was impossible to hunt, stories were told and the shaman was busy healing the sick and appeasing Sedna—also called Nerrivik, the goddess of the sea—if she was angry and withholding marine mammals. The shaman traveled through mountains and under the ice to the bottom of the sea to entreat the powerful spirits there to release animals to the hunters and to ask for good ice and good weather.

In earlier years the qasig was the domain of men. No women entered except to bring food during dances. Later, the ceremonial house doubled as a guest and community room where walrus hides were split and sewn, ceremonial masks were made, walrus ivory was carved, and the Bear Dance was held. "Enormous quantities of meat were cooked and taken to the qasig on large wooden platters," Charlie Johnson, an Inuit polar bear biologist from Nome, wrote. "The hunter sat in the middle of the room with the bear's head before him, as people feasted. An improvised song was sung about his courage and the bear's ferocity. The balloon of a seal's bladder was burst to mark the hunter's turn to dance. He leapt and shouted as if in a struggle with the bear. His wife danced with him. When she became tired, his mother took his wife's place, then an aunt, until everyone wore out. Then the bear's skull was thrown into the sea."

Joe remembered celebrations that lasted a week. "Time was bigger then," he said. Time was elastic, measured by migrations and seasons, light and dark, dancing and rest, hunger and satiation. Before they lived by the white man's clock, inserted into the culture by missionaries to remind people to come to church on Sunday, the minute and hour hand didn't exist. No calendars hung on the kitchen wall. Time was told seasonally, by snow and sun and the arrival and disappearance of bowhead whales.

Joe and I are still walking. The wind begins to howl as Joe scrutinizes the tiny, caved-in houses. He looks puzzled. "I don't know which one was mine," he says solemnly. "Did I ever really live here?" he asks. "This is the dilemma of modern man facing backwards. Can we go back? Where is forward?"

A Bering Air plane lands. It's possible to order a pizza in Nome and have it delivered in Wales when the plane makes its daily scheduled run. Yet there's almost no food on the shelves of the general store. When I ask why, Joe shrugs. "I guess the people running it aren't doing a very good job. Didn't pay their bills. Now the wholesalers won't deliver food."

Joe sees the scarcity of store-bought food as part of the general poverty that infects the villages of the Seward Peninsula. "Despite the wealth brought in by the oil companies, this village is below poverty level," he says. "No insulation in the houses, no food in the stores, and no flush toilets. We still use honey buckets like always." Material wealth is a kind of poverty. "We don't need white man's crap," Joe says. "So it's important to redefine poverty. For example: It's true that in the old days, before white men arrived, we had no doctors and not much in the way of formal education, but we were rich in food and imagination, and something else we don't see much of, which is gratitude."

The people of Wales invented and made everything they needed, Joe tells me. They made kayaks, umiat, harpoons, bows and arrows, soapstone stoves for heat and light; used skins for clothing and shelter; had dogs to help transport their things from camp to camp; and along the way told stories, sang songs, and made carvings. "We hunted and were hunted by weather and spirits and polar bears. Accidents occurred and death was frequent. But we expected nothing less. We weren't looking for salvation. We just lived."

Late in the afternoon a young woman, Metrona, comes to visit Joe. Inuit people still have famine in their stomachs. They have memories and stories like the one from Barrow about the girl and her brother who were eaten by wolves and subsequently became caribou. "We were warriors and we were cannibals," Joe reminds us, smiling.

Metrona: "You could say that we Inuit are always looking for something to eat. You might see a pretty hillside: We see it as a place to look for berries."

In her 30s, she's bright and self-possessed and not one to pass up an opportunity to make a little money. She has come to sell us her crocheted hot pads, mittens, and caps.

"Not many people live here," she says, her eyes sparkling, "but it's home. I can't wait for summertime. We pick a lot of greens. When they're real young, wild potatoes, onions. *Shusha* greens, *eviks,* sour docks, *putnuqs, koonaliks, evalunks,* and many more. Those are just the ones that grow here. There are others on Little Diomede. We preserve them in seal oil in big glass jars. They stay good all winter. We also pick a lot of berries, cranberries, blackberries, blueberries, and salmonberries, and make jam. We pick them up on the mountain and go over to the other side. But last winter, the weather was strange. It rained a lot and froze some, and the flowers blew away and the leaves froze, so we didn't get many."

She shows me her arms. "I got my tattoos here. A friend did it. In the old days, women in Wales had their chins tattooed, using the soot from the blubber lamp. On St. Lawrence Island the tattoos went all over the face, arm, and shoulder. I try to remember all of what my grandparents told me. Like eating little orange snails raw to prevent getting cross-eyed. They taught me that when the wind blows northwest, the ocean color changes to brown, and that's what brings the clams and snails in. We are always watching the weather and how it can bring us food, because starvation is still real."

Joe's stomach growls and he suggests we see if anyone in the village is cooking. We tromp up a snowy hill to visit Betty, who, at age 82, lives alone in a clapboard house at the top of a hill. She's small and soft-spoken, and so shy at first that we sit in silence for a long time.

Outside, wind winnows a snowdrift into an elbow that leans against a shed. An empty bird's nest rests in the eave. Finally Betty smiles. "A long time ago we were strong and fast young people. Now I'm not doing so well. Back then we were healthier. We had only the Eskimo diet and greens."

Betty came from a reindeer-herding family. Christian missionaries brought domestic reindeer to Alaska in the late 1800s. Men who had been marine mammal hunters for something like 20,000 years—the exact date is still contested—were being "retrained" to herd domestic reindeer. The training came with a hearty dose of Christianity.

"You could get a herd of reindeer if you converted, or put another way, food came with religion," Joe says. "It was a clever idea, but not clever enough." The reindeer project failed, was restarted, and failed again. But religion stayed.

Betty's father was one of those retrained men. He learned to herd and managed reindeer for the resident missionaries, the Lopps. "We lived in a sod house with our in-laws. Then my husband built this house. It's almost ready to fall down now. He built that birdhouse so I could watch the snow buntings come and go in the spring. We didn't go whaling much. We just had the reindeer. My mother wouldn't let us go to the village celebrations or visit the shamans. Her religion said those dances in the qasig were evil. But when I was young, I went anyway. There was nothing that made me afraid."

Except hunger. Reindeer were introduced to alleviate famine. Everyone had famine stories. Between 1881 and 1883, the people along the coast from Wales to Point Hope had very little game or fish, and many people starved. The famine was said to have been caused by shamans fighting among themselves. People from the regions around Noatak River, Kotzebue Sound, and especially Kivalina were hard-hit. The caribou stayed away. The seals did not haul out. The fish did not spawn in the rivers. Geese and swans went elsewhere. One villager said half the population of these areas died and the survivors moved to Barrow, Unalakleet, and the central Kobuk.

The crew of the ship *Corwin* found dead people "scattered about the banks of a stream near Cape Thompson." An epidemic of flu followed soon after, brought by the whites. The autonomous nation of the Kivallinigmiut, northeast of Wales, ceased to exist altogether.

Islands were particularly susceptible to hunger. If the seals and walruses and bowheads didn't come by, and the ice was bad, there was nothing islanders could do. When hunger struck one village, its people packed up and moved to a village on the other side. Food was always shared, even if it meant eventual starvation for everyone.

James Aningayou of St. Lawrence Island remembers hunger: "We had short of meat, and poor year, poor spring. We had a little meat from the spring hunting, so we had used up during the summer. Then in the fall we have nothing to eat while waiting for the ice to appear. My stepfather had two dogs, I think. He kill one, is very fat. Then he boil the muscle of the hind legs and front legs. That was good."

A villager said: "The best hunting was when the ice first got here. When it started coming, everybody would go to the top of the mountain, they were so glad to see the ice coming in. They had been eating old meat for a while, some families were out of meat, they had been along the beach all the time looking for seaweed so they would have something fresh to eat. Everybody wanted the ice. It meant they would have food."

COMING DOWN FROM the hill at the far end of Wales, we pass a man on a side hill in front of his house. His snow machine is in pieces on the snow. He'd tried to get to Shishmaref but ran out of gas. "Walked home. Pretty cold with that wind. Now I'm trying to get this old snow-go going." He offers to sell Joe a carving. He needs the money to buy gas. It's a fossilized piece of mammoth ivory etched with images of ice age animals. Joe turns it over and over carefully. "You're not asking enough for this. You should sell it in town, in Anchorage. You'd get a lot for it."

The man protests. "Anchorage is a long way away," he says. Joe gives him a hundred dollars in cash but refuses to take the carving. We slide down a steep hill in deep snow to the high school. It's a modern building with central heating and flush toilets. The classrooms are equipped with computers; moviemaking equipment; art rooms for painting, carving, and woodworking; and a biology lab. Two enthusiastic high school girls ask if they can film Joe and me at the end of the day. "Are you doing oral histories in your community?" I ask. They shrug.

Between classes the halls fill with students of all ages. The mood is high-spirited and friendly, with students, teachers, janitors, and administrators intermingling easily. Ray's son Clifford, who works as a school janitor and unofficial counselor, waves. He's busy showing a young boy how to hold a carving tool. Joe looks in and nods approvingly.

I'm snagged by a teacher to say hi to her first graders. No rows of desks for these kids. The class dynamics are free-form and enthusiastic. The kids fire hundreds of questions, and two older girls interview us using a new video camera on a tripod.

"But no one speaks Inupiat, our language," Joe says as we put on our parkas. It's 50 below zero outside with the wind chill and getting dark. "They are using someone else's language and they don't even know it. English doesn't have the words to explain who we are, what we know about the land and ice, and how the ice is changing."

One of the young teachers invites us for dinner. She's a gregarious Vermonter who has lived in the North for years. Her house is modern and cozy. Her ponytailed Inuit boyfriend, from Little Diomede, is carving a walrus-ivory handle in the shape of a polar bear for an ulu. It's exquisite, and Joe compliments him on his carving. "My father thought that with every generation, things were being done less well. But seeing this, I'm not so sure," he says. For dinner, we have a reheated tuna-noodle casserole and tea brewed from local herbs.

They lend us a flashlight for our walk home in the dark. The old part of town where Joe grew up is covered with hoarfrost. He shines a dim light on the broken boards. "We were never cold," he says. "We always had fires going and food on the stove. Mountains of food—ducks, walrus, and seal meat. They'd divide it all up so everyone had something to eat whether their hunt was successful or not," he says, switching off the flashlight. The ruined houses shine in the night.

JOE KEEPS SAYING we're losing daylight, but all I see is the night sky wiped clean by the white cloth of snow. Maybe what he means is that his grasp of who he was and who he is in Wales is elusive. The whiteout is nearly continuous, with only short glimpses of the revolving airport light or the headlights of a snow machine dashing by. A wind picks up and gusts hard, shaking the windows. Earlier in the week five polar bears were sighted nearby.

Above the sky is darkness, and below the advancing and retreating ice pack counterbalances dusky days. When the ice came down the strait from the north, *nanoq,* the polar bear, was often seen hitching a ride. Bears travel, hunt ringed seals from their white deck of drift ice, eat prey, beachcomb, grab eider ducks, and wander the land between Tin City, where Joe's father once worked, and Wales.

Villagers don't want polar bears too close, yet they are always watching them, learning from them. They found a den carved in a wind-hardened snowdrift at the second inlet to the Lopp Lagoon near town. When spring came, the bears swam across gaping leads in the ice. Between March and May they began moving north again, hitching rides on pack ice as it receded from the shore. Another hunter found a denning female on the east side of Little Diomede Island that had excavated a cave in the cliff 50 feet above the shore. It was 4 feet high, with two chambers each 13 feet long. Only females hibernate. Once they go in a den, the males have been observed blocking the entrance with hardened snow, then going off until spring.

A hunter from St. Lawrence Island said he once had trouble with too many bears coming near, so he lit a bonfire using bear fat to ignite it. The smell scared the bears. They dived into the ocean and began swimming. Others joined them, forming a wedge in the dark sea that he said "looked like a large white ice floe."

In the morning there are no bears in sight. Ronnie, Winton Weyapuk, and others come to work in the office, a small room off the main hall. We talk to Winton about uncontrolled natural resource extraction, about sovereignty, self-rule, sustainable economies, and indigenous peoples—issues that are always on the front burner up here, but about a hundred years too late, Joe says.

Winton explains that the 1971 Alaska Claims Settlement Act provided the 80,000 Alaskan natives with 44 million acres of land and $962 million as payment for lands given up. But the question of how to use this "tool of money" for the benefit and health of the land and its people is a continual debate. Still, spend it they do.

Oil and gas development have been intense and have brought great wealth to the North Slope Borough in Barrow. Drilling on land and sending the oil south through pipelines bother the

Alaskan natives less than the idea of the offshore oil drilling being proposed by Shell Oil. These people are maritime hunters, and the impact on marine mammals would be crushing. Oil spills from tankers have shown the world that petroleum extraction is environmentally perilous no matter how you move the crude oil around.

North of Point Hope, where the Bering Sea meets the Beaufort, Shell Oil has been issued an exploration permit to look for oil using seismic testing that, in itself, endangers bowhead whales, walruses, birds, and fish in those waters. As the pace of global oil grabs accelerates, environmental impacts are seldom thoroughly assessed. Now, after a lawsuit put forward by Earth Justice on behalf of the Inuit hunters of northern Alaska, the Ninth Circuit Court of Appeals has halted Shell Oil. But it is only a delay. It now must decide whether the potential for environmental damage was properly considered by the federal agency, the Department of Interior, when it issued the exploration permit to the oil company.

"The climate is our biggest problem," Winton says. He's aware of some of the complexities of the ocean-atmosphere exchange and its effect on the warming or cooling of the planet. "Maybe whatever we do will be wrong because it's not been going in a good way for a long time. If it gets hotter, it will be bad; if it got real cold, there would be no cracks in the ice and all the marine mammals would die."

He's been keeping weekly sea-ice observations in an attempt to chart the course of global warming on the Bering Strait. He hands me a sheaf of paper with his writings:

*January 8, 2007:* Winds are calm. People reported that they heard the ice piling on the pressure ridges. One resident said he heard it before midnight and it sounded like a jet engine, a continuous loud roar.

*January 10, 2007:* Strong storm overnight with winds from the SE gusting to 50 mph. Snow and blowing snow.

*January 12, 2007:* Winds decreased to 10 to 15 mph from the SE. The pressure ridges look as if they have been sheared off and a new row of ridges formed closer to shore. Three seals can be seen on top of the smooth ice three-quarters mile from the village.

*January 19, 2007:* The winds have been from the NE for four days. There is extensive open water beyond the shore ice.

*January 22, 2007:* The pressure ridges along the edge of the shore ice are about a half mile farther out than usual. We have not yet ventured out to the edge to examine their structure. They look fairly high. Hunters usually wait until they are certain that the ice is safe to travel on and will not break off and carry them away. It has changed to a near solid white color, which indicates it is safe enough for travel.

Winton takes me for a quick drive to the shore on his snowmobile. The frigid air feels good on my face. There are cracks in the ice but no open leads. The sky has cleared, and far out I can see a white wall of pack ice and beyond a vague blue hump that is Little Diomede Island and Siberia—where Winton's relatives originated thousands of years ago. He says his favorite sound is wind-driven pack ice piling up. I hold my hands behind my ears and listen: Blowing snow scratches, a tilted ramp of ice gives out a hollow crack as the tide goes out, but that's all. No primordial collisions.

Ice never sleeps. It has its own life. It is always moving. Maybe our ideas about time and consciousness come from sea ice. The way it piles up, pulls apart, melts, and freezes, spins and splinters into drifting islands carried by tidal currents to Siberia. The way its albedo drives heat away yet also sequesters black soot that causes it to melt and alters what climatologists call the surface energy budget.

Winton, like all of us, tries to understand exactly what is happening. But if the consensus-driven climate models from the scientists at the Intergovernmental Panel on Climate Change (IPCC) have failed to give us accurate predictions, then where do we look? Winton says there's nothing for him to do except observe. "We don't have theories about ice and weather and climate. We have experience," he says.

I ask him to describe the ice we are seeing: "*Tugayak* means 'shore ice breaking up and separating from the beach—ice going away.' " He has to yell over the wind. "Now the *ivu,* the pressure ridge, comes in fast and gets shoved up on the beach when the wind is strong like it is now, and the tide is high. Pressure ridges form where it just starts to get shallow. They aren't as big as they used to be and they break up more easily. The pressure ridges act as a barrier allowing the ice to stay. It makes our world calm. It gives us nice weather. Wales was a place where winter once came easily. Now it's stormier. Now winter may not come at all, or maybe too much of it will come. Either way, I think we are not going to prosper."

Back in the multi, Joe, Winton, and I scrounge around in the nearly empty cupboards for tea bags and cookies. Even a brief foray out into the deep cold brings on hunger. I ask Winton how many generations of Weyapuks have lived in Wales. He says it goes too far back to count. "For as long as humans have lived on the Bering Strait. Maybe 10,000 years or more." He was brought into the world by a village midwife, as those before him were.

"The women helped each other. Now they send mothers-to-be to Nome a month before the delivery date. The village doesn't participate anymore, and the women come home with little strangers in their arms."

There's a scratching sound on the side of the building, and we go to the window to see if it's a bear. In Kivalina, one of the villages now slated to be moved because of coastal erosion, a young man tried to save his pregnant wife from a polar bear. "The bear attacked her" Winton says, "and the husband got the bear to chase him and let go of his wife. He was killed by that bear. The wife and baby live still."

Joe and Winton's conversations go round and round as if driven by a circular wind, from talk of the old days in Wales to the new poverty of modern times to climate and weather.

Winton says that in the last years a lot of beach has eroded, at least a hundred feet since 1990. "There used to be two rows of sand dunes, and both of them got washed away," he says. The first row of houses is now in danger. They're built on the beach, and if big storms keep coming, the way they have been, those houses will wash away.

"Maybe a seawall will be built. I don't know if that will help. Up the coast Shishmaref and Kivalina will soon have to move because of coastal erosion. There are more storms, worse storms, moving in more often. And now we have the warming ocean going against us. Wait until the sea level rises. We'll all have to move."

"The hunter follows the ice," Winton says. "In the spring we look for a smooth, low place to chop a trail through the pressure ice. We use pick axes and shovels and chop off clumps of ice and smooth them out to make a place to drag our boats out and wait for the whales. We do this in mid-March before whaling season begins."

This year, he tells me, the pressure ridge was out farther than normal, and since 2000 the whales have gone by much faster. "The leads in the ice used to close up, and that temporarily halted the migration. Now they are open all the time. We used to see whales for a month; now it's down to two weeks. Same way with the walrus. They used to get up on ice floes that moved north for a whole month. Now that kind of ice lasts for only a few weeks, then it's gone."

Summers they hunted geese and ducks and fished for salmon at the mouths of small rivers. In late August they went north of Shishmaref to hunt caribou. "In the old days we walked up the coast. Everyone had summer camps there. The women picked greens and preserved them in seal oil. Then we waited for the ice to come in."

Now the inland hunters from farther south tell Winton that there are lakes drying up, landslides along the rivers, and no berries. Warm winds come all the time from the west and the willows are bigger. There are no muskrats, no beavers, and ice collects on the caribou's feet and makes it hard for them to travel. Like walking on broken glass.

"These days there is no more multi-year ice. No old ice at all," Winton says. "That's what we all depended on for water because all the salt had percolated out. We started losing it maybe 10 or 15 years ago. That's how long ago things like wind and ice began to go fast."

The Arctic is always changing. Twenty-three million years ago Alaska had the same climate as Pennsylvania today—the very same species of trees. A mass extinction of ice age animals began 15,000 years ago. Grasslands turned into bogs and grazing animals starved. Now the interglacial paradise in which we've been living is coming to an end as human-caused climate change escalates.

While we should be headed for a new ice age as determined by Earth's orbital cycles, the level of $CO_2$ and methane emissions and the heating they are causing is overriding the natural cooling trend. A new wave of mass extinctions will surely come.


JOE AND I VISIT the new houses at the shore. They're roomy, but none have proper insulation or hurricane-proof stabilization. People complain about heating bills. Many, like Pete and Lena Sereadlook, can't afford a phone. A wind turbine that could generate cheap energy for the whole village stands motionless. I ask why it's not running. No one knows. "It's owned by a Kotzebue energy company. It hasn't worked for a while."

Pete and Lena live so close to the water it's possible that the pack ice could come through their front window. Lena is feisty and tomboyish; Pete is older, frail, soft-spoken. Joe and I have come to look at the footage Pete has been shooting for 40 years, but it's all on 8mm and his camera is broken, so there's nothing to see.

He says he was born in a sod house in the old part of the village. "It was real good. We played outside all year round. We skated and played football with a ball made of reindeer skin. We made it ourselves. You don't need to buy much. We had a big dog-sled and seven dogs. I had my own windmill out there. It worked good, not like these new turbines. They aren't even running. I guess they're here for decoration! We had gas lamps and wood-stoves, no toys. We used cans and rocks for toys. They work just as good. Dad made our skates from the frame of a steel bed. We tied them on over our mukluks. We didn't believe in going to church. Whenever we heard the church bell ring, we ran away and skated out on the frozen ponds or else played Eskimo baseball. There are still two sod houses 15 miles up the beach at Sinauraq. That's where my parents were born."

Pete says there were lots of whales and walruses back then. "We listened for them. You could hear them in the spring when they started coming up a lead. Now what you hear is snowmobiles."

Joe and I walk along the shoreline. There are skin boats, kayaks, and "aluminums" on racks. A yellowish hump at the shore looks like a pouting lip, but it's only rotting ice. The bowhead whale hunters divided themselves up into family and working units headed by captains. There are four in Wales. One of them, Frank O., comes by. Young and fast talking, he grew up learning weather and ice from his father and uncle. "Now the barometer rises and falls too fast. And our seasonal storms, the ones that had winds of 30 to 50 mph, are 90 mph," he says.

His eyes narrow as he looks across the Bering Strait. "Our polar bears' homeland is melting," he says. "Can you believe it? This year we've had only one bear visit. Usually they're all over the place. Eight or nine of them. I don't know what's going to become of them, or of us," he says, taking off his mittens and rubbing his hands. They are chapped and meaty.

"I grew up hunting every day," he says, looking out the window. "Now we're a lazy people because we've got snow machines and four-wheelers. We used to pack ice from places by the mountain for drinking water, dragging big chunks on our sleds. There are only a few dog teams now. They are getting fewer and fewer. Subsistence hunting is expensive. How can it be subsistence if it costs money? It doesn't make any sense.

"We've lost our language. The white people took it away from us, but we haven't tried to get it back. That's why our story is the way it is now. If you don't know the words that describe the weather, ice, and animals, then you don't know the life. That's why some of these kids are lost. They don't know their way around here. They don't even think about the things that make up my whole world."

FRIDAY. Pull-tab night. The wind howls. A villager named Gabriel comes into the multi and unlocks a door to a tiny room I thought was a broom closet. Instantly, it turns into a small shop. A counter folds down, and behind are shelves lined with boxes of "pull-tabs," sodas, candy, and peanuts. I have to ask what pull-tabs are. Joe says they are the most passive form of gambling imaginable. "You pull the tab like opening a can of pop. Underneath is a printed number, the amount of money you win, if you win any. That's the amount of money you are paid. Pull-tabs cost a buck apiece. Usually there's a blank under the tab, so it's almost impossible to make a profit."

Gabriel switches on a small TV hidden in the corner and puts on the sci-fi channel. The blaze of the screen turns his brown face blue. A tall teenage boy comes in clutching a wad of dollar bills. Forty dollars in all, he tells us, throwing the bills down on the counter as if on the bar in a Western movie. Instead of whiskey, Gabriel doles out 40 bucks' worth of pull-tabs. One by one the young man, sitting alone in the glare of the TV, opens them. Five are marked one dollar; the boy gets five dollars but loses thirty-five.

Clifford sticks his head in to see if anything is happening. He wears thick glasses and Carhartt overalls. "Friday nights are pretty dead," he says. I ask what happens on Saturdays. Muffled laughter. "There used to be storytelling, but TV wiped that out. When I have a story to tell, I put it into a carving, but my sight is going. It's almost creamy. They're going to put a laser on it. In Anchorage. As soon as I get the money to go," he says.

"A lot of things are changed since we were kids," he tells Joe. "Like language. Remember how different each village dialect sounded? Barrow dialect was musical. Shishmaref was real slow." He says he started hunting when he was five or six. "They put me in the center of the boat. Been watching the Earth conditions since I was little. Had to. No way else to survive. In a boat you do a lot of thinking. Winds and currents are not like they used to be. They go back and forth like they can't decide. We have a lot of erosion here, but not as bad as Shishmaref. We could sell our mountain to them. They need some ground for the village. When we were kids the ice used to shift north, then come back. Now the ice stays up there, then it blows offshore and the current collides with it. It goes over to Diomede.

"I don't hunt for a living. I'm a custodian at the school. I help keep the kids' morale up. Unofficial counselor. I live alone, so I'm here for the kids. Hopefully we can pass down some traditions and they can learn to watch the weather on their own."

Snow scratches at the windows. Wind shakes them. Then the quiet evening of pull-tabs with only one player comes to an end. Gabriel closes up shop and goes home. Alone in the big community hall again, Joe wonders if they ever have traditional dances here. "Might be more exciting than pull-tabs," he says, looking out the window. All he sees is his reflection. No outside world at all. "Did I come here just to see myself?" he asks quietly. The sky has been dark since four.

On our last night in Kingetkin, Luther Komonaseak bursts into the multi. "I really wanted to talk to you before you left," he says breathlessly. His baseball cap bears the name Tikigaq, the Inupiat name for Point Hope, the oldest continuously inhabited village in Alaska. People have been living there for at least 14,000 years.

"Up there," he says, meaning in Point Hope, "the people are more traditional than we are here. I don't know why," he begins. Sharp nosed, fast talking, and passionate about life in Wales, Luther is a 52-year-old whaling captain who says he learned whaling from Winton's father and from Ray.

He grew up learning from everyone because his dad didn't have a boat. He was sent to Little Diomede when he was nine. "They had rock houses and half doors. You had to crawl through a tunnel and come up through a door into a sod house. They taught me to hunt the traditional way there," he told me.

When he came back, he started his own crew. That was 1987. The ice was frozen all the way out, and the whales were coming back, but the ice blocked them. He didn't get his first whale until 1994. That's the kind of patience it takes to be a hunter.

"The day I got my first whale the water was boiling with bowheads. They'd stick their heads up and look around. They know when you're scared and won't come close to you anymore. So you can't have fear and provide food for your village.

"After, I had a whale feast. Three boatloads came over from Diomede. They respect me now. It took a long time. I gave a lot of my first whale away. It comes back to you in other ways. It takes three days to harvest a whale. The role of the captain is to keep track of who gets what and to be fair to everyone. A square from under the chin goes to the harpooner. A big square of the stomach goes to everyone. From the bellybutton to the front and all the way around goes to the captain. Parts of the skin from the back go to the whaling crew, and the other goes to the crew that helps tow in the whale. Each community has a different way. I'm writing it all out so it will be remembered."

He says he always knows if people are having food problems, and he brings meat to them. "Last year I brought Faye some seal and walrus meat. Boy, she was happy. We have two elders at every council meeting. They are Faye and Pete now.

"We try to take care of our future by using the past. That's an unknown concept to most kids these days. We communicate with the young people who are not interested in going out hunting. We try to push them in the direction of tradition, because that's what we have here. That's all we have.

"Wales was a radical place when white men first came. Radical in the knowledge it took to survive. Wales was like a hub city before Columbus. Stories were told here, things were bartered. The Messenger Feast originated here when a whaling ship came to Port Clarence and a runner with a message on a big pole came to the ship saying that we were having a feast and they were invited.

"My grandson and son go out with me. At first the younger one said it was too cold; now he's hooked on hunting. He was only four when I first took him.

"I went up to Point Hope and hunted there. They still use skin boats, no outboard motors. The pressure ridges are two and a half miles out, they're really high. Bowhead whale hunting is much better there than here because the whales go around Point Hobson, then come by Point Hope, real close. Now everyone is having problems with the ice, and a lot of changes are happening. The Barrow people say they are seeing narwhal for the first time. That means the narwhal are making it all the way across the Northwest Passage from Greenland to Alaska." (They are normally only on the coast of Greenland and in the waters of eastern Nunavut.)

"Summers are hotter too. Our water supply went down to nothing. Now we're getting sick from the water. We have dragonflies. Never had those before. And robins, porpoise, sharks, and orca. Real strange. It'll come a time when the migrations of eider ducks and walrus and whales get so mixed up that they'll come at the wrong time and weather. The wind and the currents and the coming and going of ice will go against them.

"These days there's either too much ice or not enough. You can hardly forecast weather the Inupiat way. The elders taught us signs about how to know the weather. Like a wind cap over the mountain at the southern end of the village meant high winds coming. The clouds never lie.

"I'm on the International Whaling Commission," Luther goes on. "We have to tell people that we aren't commercial whalers. We're just feeding ourselves like we've done for 10,000 years. The oil companies neglect the subsistence whalers. We tell them they can't use big ships in the migration route. They're supposed to stop during migration, but they don't. In Kaktovik, at the top of Alaska, the whales are going further out because of the oil industry activity. A lot of things are going haywire.

"Fall time, the whales return later. Now it's the end of November, beginning of December. Bowheads go south a bit; humpbacks go to Hawaii; grays to Baja. It's not just right here that we are concerned with. What happens here with the ice affects everywhere, just like that," Luther says and snaps his fingers. As he does so, all the lights in the village go out.


OUR PLANE LIFTS OFF in a snowstorm. We head east from Wales to Nome. The pilot is young but he knows the way. It's minus 22 degrees with a light wind. To the south, down the nose of the Bering Strait, is King Island, a chunk of granite sticking up out of a tormented sea.

The King Islanders, the Aseuluk, were forced off their island by the government and had to move to Nome, 90 miles away. "So Wales people and King Island people were bunched together as two outsider groups," Joe says, looking at the island in the distance. Its sides are so steep it's hard to imagine anyone living there at all.

They say that King Island came into being when a giant fish was harpooned by a hunter who cut a hole through the fish's snout to tow it back to shore behind his kayak. A storm came up and the hunter had to leave the fish behind. The fish turned into an upended stone and became the island of Ukiovak—King Island. The name means "a place for winter." The island and the Arctic culture that arose on its vertiginous cliffs are deeply expressive of the interaction of humans and their environment—in this case a plug of rock sticking out of the ocean. Against all odds, humans thrived in this difficult and unlikely place.

At two and a half miles long and almost two thousand feet high, King Island has no harbor and no landing place but once had a hanging village of driftwood houses on stilts tied to the precipice with half-inch braided walrus-hide thongs. Each cantilevered platform held two separate houses, their frames overhung with walrus hides. "An Eskimo duplex," Joes says, jokingly.

The houses were small, about ten feet by ten feet, with a storage shed in front for kayaks, paddles, skins, harpoons, and spears. The ceilings were insulated with dried moss and the walls were stuffed with grass. "Eskimo wallpaper," as Joe called it, was sealskin. The roof was laid with split logs. Window and door coverings were made of walrus intestine that served to let in light and also functioned as lighthouses for homeward-bound hunters.

A deep cave near waterline with permafrost walls served as a cold storage for the islanders' food. Walrus ropes were used to help climb up from the cave to the houses. Women carried huge loads of meat up and down the cliff in walrus-hide backpacks.

Before a hunt, the villagers climbed on top of the dance house roof at dawn, lit a small fire, and sang as the men went out. The men's skin kayaks and umiat were often strapped together for safety. These were not ice hunters—they harpooned walruses in open water.

When meat was scarce, the men climbed bird cliffs wearing walrus-hide harnesses. "Straight up among screaming birds, they captured cormorants and stole murre eggs," Joe tells me. The residents of King Island were not unlike birds themselves, as they lived perched on cliffs in houses that shook with the winter winds. The women tanned hides with their own urine, and men made tools with whatever materials were at hand. Baleen strips were used for fishing line, odd-shaped rocks were used as hooks, water was carried inside a pouch made of walrus intestine, snow goggles were carved from driftwood, shovels were fashioned from the shoulder blades of walruses, knives were flint or jade. The flat tops of the houses were used in good weather for open-air work spaces.

December was called the Month of Drumming. When the spirit of Sila was appeased, dance competitions were held, new songs were composed, and shamans went to work keeping whatever social problems had erupted under control. There was ritual, storytelling, and dancing. Even in this tiny, hanging village, there were four dance houses on the island. Families belonged to individually named dance houses, and friendly competitions were held among them. The long winter entranceway was cold and narrow. Like a mole, a person crawled along, then popped up through a hole in the floor and entered the dance house. Driftwood beams were supported by four poles, and benches lined the walls.

Winter could be a time of hunger if the stores of food from previous seasons ran out. Extra hides were always put aside: From them, nutritious broth could be made to get people through until spring. If hunger became extreme, villagers moved in with each other to share food and warmth. Nothing was hoarded, even if it meant that everyone died.

King Islanders were maritime travelers: By kayak and umiaq they went south to St. Michael Island, north to East Cape, Siberia, northeast to Point Hope, Kotzebue, and Wales. Their dogs were weather forecasters: They were trained to sit in the bow of the boat, sniffing the air and feeling the currents, barking if there were shifts in the ice. Moving ice could crush a boat or break off and carry away an umiaq and its crew.

Joe's mentor in Nome, Paul Tiulana, was from King Island. Paul learned to listen for the haunting mating song of the bearded seal by sticking his paddle down into the water and holding it to his ear. He knew the sound of spirits as well. "If a spirit wanted to come into the house, there was a big bang on the roof. Then fog entered the room and the seal-oil lamp flickered. Medicine men and women in the village cured the sick, found lost objects, and danced for good weather. They knew when villagers were in danger out on the ice."

He recalled a story from 1949, when three men drifted out from the island. One was his cousin, whose parents, upon hearing a loud crack under their house, knew he was dead. Another man on the ice was named Ayek. His mother knew that if the man's extra pair of mukluks stopped moving back and forth during the night, he was dead. The mukluks kept swaying, and after drifting for 17 days in the Chukchi Sea, Ayek was found alive. Now, two generations later, his grandson Sylvester Ayek, an internationally recognized artist in his 50s, returns to King Island every summer to make his Calder-like sculptures.

When the prey was polar bear, there was a dance to honor the animal. Food, rawhide, and furs were given away. Both men and women composed their own songs. In any time of plenty, there were large dances to which people from such far-flung villages as Wales, 30 miles to the northeast, were invited. There were cross-dressing dances in which men dressed as women; bench dances when the entire dance was performed sitting down; and competition dances when each dance group competed with another for a prize—perhaps a dance fan made from the chin whiskers of caribou. Men and women changed mates in a dance called the Yugaisug. A man would throw a bearded-seal rope over his desired "exchange wife," go off with her for sex, and return her by morning.

A contemporary King Islander referred to one ancient source of entertainment as Eskimo TV: A large wooden dish would be filled with seawater and a gut-skin parka draped around it, using the neck hole as a "lens" to look into the water. There, they could "see" if hunters who had gone out were safe. "But now, when looking into a bowl of water, all that is seen is a tiny sparkle of light, a very tiny one. Nothing else."

WE APPROACH NOME in falling snow. No visibility until the runway lights show, the altimeter like a clock winding backward from heaven to earth, then forward again as if to reflect the jolting leaps between old and new in this ancient culture.

In the winter gloom there are Christmas tree lights, even though it's January. One young man has been "replanting" discarded Christmas trees. They stand in a single line out on the sea ice with a sign: "NOME'S NATIONAL FOREST."

In 1924, after having traveled for three and a half years by dogsled across the Northwest Passage, the Danish explorer Knud Rasmussen and his Inuit lover, Anarulunguaq, arrived in Nome from Kotzebue on a mail schooner. He wrote: "Nome lies on a grassy plain with a fine range of fertile hills in the background."

Thirty years before he arrived, the population of Nome was negligible. There were a few Inuit marine mammal hunters and their families. Then the 1900 gold rush swelled the population to 10,000 people. Rasmussen was astonished by the array of native peoples from northwestern Alaska in one place: "The entire population of King Island, the Ukiuvangmiut; the inland Eskimos from the Seward Peninsula, the Qavjasamiut; the Kingingmiut from Cape Prince of Wales; the Ungalardlermiut from Norton Sound and the mouth of the Yukon; the Siorarmiut from St. Lawrence Island; and the natives from Nunivak Island are here," he wrote in his expedition notes. The newly arrived hunters and their families lived in tents flanking both ends of town and made carvings to sell to tourists and gold rushers.

It was here that Rasmussen encountered his first taste of American racism. When he and Anarulunguaq entered a restaurant together, they were refused service. Rasmussen was shocked. They went back to their hotel, changed out of their skins, and put on what he called "white man's clothes." Finally, they were let in to eat.

Rasmussen and Anarulunguaq had a child midway through their journey from Greenland to Alaska, born on King William Island in the Canadian Arctic. Knud was married to a bright, wealthy Danish woman who helped raise funds for his many expeditions. To bring his Inuit "wife" and their child home to Copenhagen was unthinkable. Same problem for Anarulunguaq, whose husband-to-be was waiting for her on Qerqertarsuaq, an island near present-day Qaanaaq, Greenland. What must have been a heartbreaking decision was made: They would give the child to a young Inuit couple in Nome who had no children of their own.

At the Polar Café, Joe introduces himself to an older man—an Inuk, a native Alaskan—sitting alone by a window that looks out on the frozen sea. His name is Miseq, also known as Roy Tobuk.

"My name means 'wet tundra,' " he tells us. Joe invites him to eat with us.

Miseq grew up with the Athabascans. They called him Iqiliq, which means "lice people" or "Indian," because they didn't know he was an Inuk. He tells us of the old wars between the Athabascans and the Inuit, and the times they captured Inuit women and took them south to the Yukon River village, where they were forced to live the rest of their lives.

Roy moved to Nome in 1945 and worked for Alaska Airlines. Eventually the street where he lives was named after him: Tobuk Alley. We try to pay for his breakfast, but he refuses. "I'm 80 years old, but I can still pay my own way," he says. Joe tells Tobuk that he looks young for his age, and Tobuk replies: "Yeah, but you should see the inside."


MINUS 22 DEGREES. Early fog cleared by a light breeze. Arctic travel is mostly waiting. We are in Nome's one-room terminal again, waiting for a flight to the village of Shishmaref. The coast of the Seward Peninsula sticks into the Chukchi Sea like the blade of an ax. Wales is at one end of the blade, and Shishmaref is in the middle. It's absurd that we have had to fly all the way back to Nome to then fly to a village only a few miles up the coast from where we started, but our attempts to find a snowmobile ride failed.

As we walk across snowy tarmac to the small plane, the young pilot calls out: "I've got dead reindeer to the ceiling in here. Sorry, you'll have to catch the next plane." First time I've been bumped for dead reindeer, I tell Joe. We go back to town to visit Joe's aunt Esther. She greets us at the door with the pink pot holder she was knitting in her hand. Half the height of Joe, she hugs his waist.

Vigorous at 78, she moved with her family from Wales to Nome by skin boat. "No compass or nothing," she says brightly. "Just looked at the sun and got our direction that way."

She tells us she was born on a dogsled during a trip along the coast. Her family stopped at a village, but there wasn't time to get the midwife. "Mom told Dad, 'My baby wants to get out real bad.' So she let me come out, right there on the sled. The midwife came down to the ice and she cut the cord and said, 'It's a girl.' That was how I was born."

She says there were no radios then and school went only through fourth grade. "I always wanted to be at camp. I did everything with my dad, seal hunting and fishing too. In December, when the moon started to get round, before daylight had returned, our family went over there to that river called Maiqta-atugvik, meaning, 'the place where one gets up to follow,' to check and see if there are any flounder there. They used their *kagiak*, their spear. Didn't even have to lie down, just had to stick it in the water and get one. Up that river, they were big and got fancy eggs. Boy, were they good!

"When I lived in Wales, we used to see a little light. It appeared sometimes behind the middle village and traveled down and disappeared in the ocean. We saw it at the horizon, then it was gone. They don't see it anymore. Not since Christianity. There were still *anakgoks*, shamans, but we were told not to tell anyone about them, and they stopped talking to us too. Their words stopped when the lights stopped shining."

Despite persistent bouts of TB that have affected her spine, Esther's eyes brighten as she says, "Summers, I always have to go fishing. I go alone. My camp is right next to where your father's and mother's was. I still get so excited I sleep real light the night before because I want to go so bad. I get up at four in the morning.

That's how you have to do it to get fish," she says, with mischievous eyes flashing. "I get silvers. Every time. Lots of them!"

Subsistence hunting will always be part of Inuit culture; wild food is essential to the diet and represents a part of the thread, however many times broken, that binds Inuit people to place.

UNDER A QUARTER MOON with a not-very-bright sun throwing pink on snowy peaks, our plane to Shishmaref flies across an Arctic plain, leaving behind a single, pointed mountain with crenulated clouds crawling up both sides to converge at the top. Ahead, there is a cover of new snow and a lid of ice on the Bering Strait, where last week it was all open water.

Where the coast ends, there's a frozen lagoon, and the village of Shishmaref is spread laterally along an arm of land—a barrier island called Sarichef. Afloat between ocean and island, the hooked moon appears to be the only thing holding the village in place.

"Our island is getting smaller," a young man says when we arrive. By snowmobile to the house of Joe's sister-in-law, we roar through sifting frost-fall between two rows of houses built so close together I wonder if we haven't landed in a Japanese town rather than an Inupiat village. The house is neat but crowded with skins, sewing materials, and small beds for grandchildren. There's a note on the kitchen table welcoming us.

We're here in "Shish" to see the ways coastal erosion and global warming can damage a place. Earlier we'd met two young civil engineers coming to Shishmaref to "grow permafrost" using a special fabric, sand, and rocks in what is possibly a futile effort to mitigate tidal erosion as a result of the retreating ice pack. Every coastal village, town, and city in the world is now threatened by inevitable sea level rise, but these Arctic villages, as well as the small island nations in the South Pacific, are the first to feel its devastating effect.

We walk to the end of the island, where houses have fallen into the sea. Those still standing are tilted sideways. Waves crash over a wall piled high with broken shore ice. Winds off the Chukchi Sea routinely gust to 60 miles an hour, bringing in 14-foot-high waves. Every time, this tiny island loses between 10 and 38 feet of sand and earth. Now night has fallen, though technically it's day. Wind-blasted snow hangs on the remaining buildings. The temperature has plummeted to minus 50.

Coastal erosion isn't new. Earthquakes brought tsunamis. Big storm waves, like the one in 1914 that wiped out much of Wales, have shaped the Seward Peninsula. The shore at Gambell, on St. Lawrence Island, has been augmented, whereas Wales, Shishmaref, and Kivalina are losing ground. In 1899 Edward Nelson reported coastal erosion during his 1,200-mile-long dogsled journey along the west coast of Arctic Alaska, recounted in his epic *The Eskimo About Bering Strait.*

Now melting permafrost and rising sea level is the relentless signature of anthropogenic global warming. Most of Arctic Alaska is underlain by permafrost. Some 70 to 90 percent of its tundra will vanish within a hundred years. As a result, there are drunken forests—trees that are sinking and leaning sideways—and building and road slumps. What Nelson called Eskimo villages—a row of simply constructed houses, drying racks, kayaks and umiat, and little else—now consist of up-to-date runways, airports, weather stations, schools, gymnasiums, and modular houses, and the cost of moving such villages is astronomical. And if the permafrost goes out quickly, there will be no roadway sturdy enough to move anything.

"Where my mother's house used to stand is now only sea," says John, a handsome and sophisticated man who teaches traditional arts at the local school. "I came back here to live in 'Shish' after college. Had to learn to hunt all over again, the hard way. I didn't know the currents and weather signs, but we had good, thick ice then, all the way past the Fourth of July." He's married now, with five children and one adopted child, and he hunts for his food while his wife sews traditional clothing.

He says that the bad storms in 1974 sent waves through the end of town all the way to the lagoon. Fifty feet of the island was lost. It seemed unusual. Now they lose 15 to 20 feet each season. High water comes with south winds, but now, no matter which way the wind blows, huge swells come in and crash onto shore. "We'll have to move soon. The seawall is sinking. The cement blocks we've stacked up have disappeared. We don't have good drinking water anymore. There are a lot of sewage overflows. Our houses are terrible. How can it cost millions of dollars to move a village when we live at the poverty level? Our house is a module with no insulation, vents that don't work, no running water, and lots of mold."

After coffee we make our way through a howling wind to the basement of the town hall, where Tony Weyiouanna has been holding court with the media. He has made Shishmaref global warming's "poster child" town. Sitting at a large desk and wearing a baseball cap pulled down low, Tony is manning the phones with a salesman's energy and demeanor and the urgent appeal of a priest. He founded the Erosion and Relocation Coalition, which has put Shishmaref in the news.

"I decided to take our erosion problems national, so I went to TV. We live in Third World conditions but get no attention. We wanted people to see what global warming was doing to us up here." Last week he was on *Oprah*. Now ABC and CBS news crews want to come. "Fifty or sixty years ago it would have been easy to relocate. Rebuild the sod houses and cart the umiat and kayaks farther up the hill behind the lagoon. But Shishmaref's relocation is going to cost $160 [million] to $180 million. We have to move the school, the clinic, the store, the houses, and the airport with a landing strip big enough for a private jet," he tells us.

"So I'm on the phone all over the world trying to raise the dough," he says, swiveling in his big black chair to answer the ringing phone.

An elder walks into the office and sits at his desk. His face is deeply lined, his hair whitened by rime ice. He takes off his bearded sealskin hat with earflaps and shakes snow on the floor. He listens as Tony's voice rises as he tries to get a donation over the phone, then tells me that the website for the Erosion and Relocation Coalition accepts contributions via PayPal.

Inevitably, this being a small village, we hear gossip about the relocation plan. "They're hoping to make ice roads to move the buildings. But now there's an argument about the new site. One was chosen about five miles from here because there's a channel in the river where a barge could come in. But the permafrost is melting so fast that the place might cave in, so another site was chosen, up Tim Creek. But not everyone wants to move there. I don't think we're going to relocate for a while. I'm going to a conference with people from another village, who are moving on their own. Maybe we'll see how they do it grassroots style."

The population in Shish is 560, with a median age of 24 years. "If they don't move soon, what are the people going to do? Where will they stand? There won't be any island left," Joe says. But Tony talks about "cultural dexterity" and the still living concept of *nunaqatgiitch*—people related to each other through possession of the land. He assures us that despite the lack of funds and local disputes about where, exactly, to relocate, it will happen soon.

"Have we been brainwashed into thinking we have unsolvable problems?" Joe asks the older man who came into the office while we were talking and is looking out the window. He turns to Joe: "What have we done to ourselves?" he asks. "See how easy the ice moves? That's like we used to be. Now we're stuck. No longer hunters of sea mammals on the moving ice."

Outside the wind has died down, but the cold will not relinquish its grip. We pass the graveyard at the highest point on the island, perhaps 50 feet above sea level, and the thaw ponds where people have being getting their drinking water. On the other side of the lagoon is a forested hill where the village might stand one day soon. How does it feel to leave one's "island in the stream," I keep wondering and remember the words of a Shishmaref woman named Hattie: "The Earth possessed us," she said.

THERE ISN'T A SPECK of coast north or south of here that Harvey Pootoogooluk doesn't know intimately—the old names and the forgotten ones and the places still used for seasonal fishing and hunting. "When young people use traditional knowledge, they shorten it. It's not like it used to be, so they know less and less. And the weather has changed, and the way of life." He has a bulging belly and a wonderfully expressive face. On a brighter note he adds, "But the capacity of the radio weather forecasters has gotten better. Our way was to look at conditions and the coming of the west wind or east wind. When the tops of the hills were covered with clouds, we were forewarned of a north wind. Clouds would billow up from the hills when a south wind was on the way. A north wind meant we didn't dare go out. But a northeast wind opened the ice, so we could go way out and get bearded seal."

Harvey's wife stirs a pot of soup and puts white bread, peanut butter, jam, tea, coffee, and cookies on the already crowded kitchen table. Their adopted grandson, who is 16, comes in and goes straight to his room. He's tall, lanky, and silent. "A good boy," Harvey says. "Wants to stay here and be a carver."

Harvey's wife lays out a series of pills on the plastic tablecloth and a glass of water. He swallows the pills obediently and returns to his stories as the young boy sweeps by and runs out of the house. "Our method of propelling a boat? Oars. Now the young kids use motors, but the cost of gas is prohibitive, so they just don't go out hunting as much. Because of gas, because of machines, they got to have a job. Wintertime they use snowmobiles. They cost a lot. But they are using them for nonproductive purposes too. Just play around. Too much of that. I caution young people not to use them except for hunting.

"Old days we used sealskin pants all the time, and mukluks. Used gunnysack squares folded around the foot for socks. I never go to school. I go third grade. My dad came from Wales. He herded reindeer. My uncle got the reindeer from Siberia and brought them over on a barge. My parents, both of them, they died in that flu. Pretty hard to live. No one to take care of us. That's why I lived with some old people. They took me because their son died.

"We used to hunt ugruk with dogsleds. I had seven dogs. Didn't see the thin ice one day and the dogs turned, but too late. My lead dog spread his front legs and tried to save me. I grabbed the sled, but it was going straight down. I got hold of the back and got out of the water. A hunter saw me and came and gave me his warm clothes."

He rubs his belly under his white T-shirt, smiling a strange crooked smile. The house is a sweltering 80 degrees, as if to make up for falling through the ice. Rummaging through a shoebox, Harvey pulls out a small driftwood carving. It's a tiny mask carved with slit eyes, big cheeks, and a crooked smile just like his. "Did you make it?" I ask. He says yes.

When I ask if I can take a photograph, he holds the mask to his chest and looks into the camera: a face below a face, both the same. Man and mask. Which is which? He hands it to me with a smile.

Inuk made masks that represented the north wind, storms, walruses, seals, and all kinds of seabirds whose inua—soul—showed through. They could represent the spirit helpers of hunters or shamans. The masks had lives of their own.

In a culture where all borders are permeable, the masks went beyond representation: They became the powerful spirits they represented, and in turn, those wearing them took on the qualities of the mask. Masks held human and spirit worlds in one face. Some masks were so powerful they adhered to the dancer's face with no strings. Such masks were alive and were treated as such. They weren't decoration or entertainment. As power-filled beings, they were concealed from view before a dance and destroyed immediately after.

One woman, wearing a seal mask in a dance, was said by the viewers to become a seal. When she pushed the mask back on her head, the soul of the seal showed itself. She sank down through the floor as they were watching, as if into the sea, then rose up again, not a human at all but a seal. Looking at Harvey's mask, carved from a piece of yellowish driftwood, I wonder if it represents some aspect of the man I'm unable to see.

Afternoon. The wind picks up as the temperature drops and darkness descends. Streetlamps line the village lanes. A gauzy light shines through blowing snow. Back at the house of the art teacher, John, we eat traditional seal and rice soup and talk about the ways in which these places are inhabited by animal spirits.

"Some bears are special," he says. "If a polar bear doesn't stop and look at you when you approach, you can tell that bear what to do. I did that once. I told the bear to come back from the shore toward me and he did. I shot at him but missed completely. He didn't leave. I shot again and got him. He was my first polar bear. He just gave himself to me."

He tells us about Little People, dwarfs that can be found in Arctic legends around the circumpolar North. "Last spring a pilot said he saw a whole group of Little People. And last summer one of the ladies at the hospital went out to gather stinkweed to heal an open wound. Where she picked it, she saw tiny human footprints. There are still things here that no one knows."

We go back a last time to see Harvey. The night is clear and stars arch over our heads like ocean spray. Out at sea, where the pack ice should be, we see layered clouds, a front moving toward land. Bering Air called to say there's a plane coming in half an hour, and it might be the last plane for a while because of the storm. We peel out of our winter clothing and ask Harvey about the weather in Shishmaref and the changing climate.

"Long time ago," he begins, "there were periods of clear and calm weather. Now you have to wait for days like that. We always ate caribou and ptarmigan. Now, too much pizza and people playing with their money. Playing bingo and even worse, pulling those pull-tabs. Gambling no good. Five times a week here. Too much!

"Used to be our land went way out. Now, it washes away, the waves cut under our land real deep. No more shallow ports. Springtime, the earth of our island is falling down. The land where we live and all that we know is soon all gone."

WE LEAVE SHISHMAREF in a light snow and fly back to Nome in a twin-engine Navajo. Once there we buy a nine-dollar bottle of Evian because our stomachs are bad from the contaminated Shishmaref water. Later, at the Polar Café we look for Roy Tobuk, but he's not there. Then we hear that he died of a stroke while we were away. The café owner tells us, "He was alone in his house down the alley. We found him there when he didn't come in for breakfast."

We walk the streets a little dazed by the news. The death of friends, the death of language and lifeways, and the death of ice haunt us in the same way. Our ignorance of how our own bodies work, and when death might befall us, carries over to our comprehension of Gaia and the planet's intertwined living systems. We busy ourselves with blinkered specialties and forget to look at the whole, or at the spirit force behind the mask. We fear the unknowable, and so edit what we can understand about complex feedback loops to the kind of consensual stories we can bear to hear.

The lies we tell about ourselves, about others and "otherness," in this case about indigenous Arctic people, are the same ones we tell about the Earth. We have no idea what is really going on in this changing climate, but we insist on doing nothing until there is a crisis. As things worsen, I almost feel relief: We have finally been put in our place. The oceans will acidify and go a deadly bright blue, riverbeds will dry up, and there will be nothing green on the ground.

In 1924, when Knud Rasmussen was about to leave Nome at the end of his epic journey, he met an old man on Main Street named Najagneq, a shaman from Nunivak Island. Rasmussen described him as having "little piercing eyes that glared wildly around" and a bandage wrapped around his jaw. He had killed several people and had just been released from a year of solitary confinement in the Nome jail. He claimed that while there, he had been killed ten times but that his helping spirits, ten white horses, saved his life ten times.

The old shaman liked Rasmussen and described the visions he'd had. Rasmussen asked, "What does man consist of?" The shaman replied, "Of the body; that which you see; the name, which is inherited from one dead; and then of something more, a mysterious spirit which gives life, shape, and appearance to all that lives."

Rasmussen: "What do you think of the way men live?"

Najagneq: "They live brokenly, mingling all things together; weakly, because they cannot do one thing at a time."

# LIVING WITH REINDEER

&raquo;&raquo;&raquo; — &laquo;&laquo;&laquo;

## THE KOMI OF NORTHWESTERN RUSSIA

*"All that lives exists. The lamp walks around. The walls of the house have voices of their own. . . . The antlers lying on the tombs arise at night and walk in procession around the mounds, while the deceased get up and visit the living."*
—Chukchi saying

ON A WINTRY APRIL DAY just south of the Barents Sea in the eco-tone between the taiga forests and tundra of northwestern Russia, we came unexpectedly upon a camp of nomadic Komi people and their 2,500 reindeer. They were packed and ready to move with the herd as the female reindeer began calving. A two-month-long spring migration would take them northeast across melting tundra, roaring rivers, and cranberry bogs to a relatively mosquito-free mountain by the end of June. The month of September, when snow began to fall, would move them down again, their yearly travels describing a lopsided circle cut through by north-flowing rivers and the flat maze of tundra, hummocks, and waterways that is their constant horizon.

The cavernous helicopter that had carried the four of us—photographer Gordon Wiltse, Inuit filmmaker Andrew Okpeaha MacLean, Russian biologist and translator Andrei Volkov, and me—had flown northeast from the city of Arkhangel'sk over a wide, factory-lined river that bent back on itself, twisting almost in half like a mind that has been lost. Just to the south was the hardscrabble collective farm where the poet Joseph Brodsky was sent in 1964 by the Communist regime to do hard labor. In his desolation he wrote: "Life steps back on itself / And stares astonished at its own / hissing and roaring forms." Behind us were the Solovetskiye Islands, ordained by Communist leader Vladimir Lenin as the Solovki Special Purpose Camp. Once an Orthodox monastery, it became one of the first corrective labor camps of the gulag system, made famous in Russian novelist Alexander Solzhenitsyn's *Gulag Archipelago.*

Russia's long history of human misery—the brutal conquest of Siberia, the terror of Stalin, the two World Wars, and the Cold War, with its persistent sweeps of radioactivity from the nuclear test sites in the Ural Mountains and on the islands of Novaya Zemlya—was chilling to contemplate as our helicopter roared over endless forests and ponds. "Here I wander in a no man's land," Brodsky wrote, "And take lease on non-existence."

The apron of tundra that stretches from northwestern Russia all the way to the Chukchi Sea in northeastern Siberia is sparsely populated. It is a vast mosaic of snow and rotting ice, snaking rivers, ice-carved lakes and ponds that from the air flash like eyes in intermittent sun.

The helicopter flight lasted four hours. We sat on hard benches facing each other, unable to converse because of the noise. A ceiling panel had come down, and electrical wires dangled over our heads. Our luggage was stacked loosely down the middle between the benches. There were no seat belts, no bathrooms. One man with a vodka hangover vomited quietly into a bag.

Green to the south, white to the north. I asked Andrei where we were. He shrugged, and said, "Somewhere east of the Kanin Peninsula and west of the Yamal." Opening a topographical map across his knees, he ironed out the wrinkles with his hand, but could not pinpoint our exact location. Out the window I saw a large oval *ostrov,* a snow-covered island of pines, which seemed to float in white haze. Beyond, a tight tangle of narrow, still frozen waterways separated spongy hummocks—tundra mats made of mosses, lichens, fungi, and berries that spread laterally, necklacing the top of the world.

Nineteen percent of Russian land is tundra and taiga forests where wild reindeer roam. The annual migratory cycle of the indigenous people of this region has been linked to herds for hundreds of years. Where the reindeer go, whether wild or domesticated, so go the hunters and herders, the only boundary being the beautiful and polluted Barents Sea.

MIDDAY. The helicopter shudders, vibrating to a standstill, then lowers onto snow. There's no village in sight, no reindeer herd, only a small group of villagers waiting with their ancient Russian snowmobiles. Michail, the head of a nearby community, greets us. He's a young, cheerful, bearish Russian. We are taken into the village of Pesha (pronounced PER-sha), where horse-drawn sleighs are the villagers' only transportation and the muddy roads are deeply rutted with slush and spilled hay.

The town is heated by coal and stinks of it. Black coal dust is mixed into the snow. Unpainted 18th-century wooden houses and barns with beautifully carved window and door frames stand side by side—elegant reminders of the Russian past and an Arctic outpost so physically remote that it seems to have been untouched since Tolstoy's time.

After a lunch of cabbage salad and meat dumplings, we climb back onto the small sleds pulled by the snowmobiles and head out to look for the herds. Reindeer blankets are thrown over us and tucked around our legs. We sputter along slowly because the machines are old and the Russian fuel is bad, and look for a group of herders, though our drivers say they're not sure where they are.

Gordon wakes me. We'd traveled for hours and I'd fallen asleep. "I think we're here," he says. Dogs bark and come running. Ahead, smoke curls up from the tops of three large tepees. Beside them harnessed reindeer patiently wait, and a group of men look up to see who is coming. As we approach, I'm shocked by what I see. These are not dark-haired, almond-eyed Nenets people, but auburn-haired, blue-eyed men and women who look Russian or Saami.

They are as surprised to see us as we are to see them. "Who are you?" I ask. They smile and ask who we are. When I say we are from National Geographic, they say they've never heard of it. Then it's their turn to explain. I had thought we were coming to stay with Nenets people. "We are not Nenets, we are not Saami, we are not Ruski. We are Komi," a man who identifies himself as the chief says. "But we are using Nenets land." Long pause. "And sometimes we marry Nenets people."

Andrei unfolds his topographic map, and the men crowd around to show us where we are. What I see is thousands of square miles of uninhabited land. "Who owns all this?" I ask. "I guess Vladimir Putin does," Piotr, the chief's older brother, says, laughing. He's tall and gaunt with curious eyes. "Yes, it must be his. There are Nenets to the east near Indiga, and Nenets to the west on the Kanin Peninsula, but here, there is no one but us, and we are not many. We have no wives, no children, and the old ones are dying out. Our fathers are dead. We are here to help our mothers." He pauses, looks around, then smiles: "In summer, the main population is mosquitoes."

The top edge of Arctic Russia is all fingers and fists sticking into the frigid sea, and since anyone can remember, there have been wild *Rangifer tarandus,* woodland caribou, here. In the 16th century there were an estimated five million wild reindeer in northern Russia. By the 1980s, that number had dwindled to a little more than two million, with another million domesticated. Now the wild deer are overtaking the tamed, and the herders say they struggle continually with wild herds leading the domestic reindeer away.

A man called Stanislav, or Stas, wearing dark glasses and a reindeer-skin tunic, hands me a piece of hard candy by way of greeting. His wide leather belt is decorated with bear teeth and bone cutouts of pine trees. A small scabbard hangs from two gold chains, and as he walks, it swings with the weight of the long knife sheathed inside. "In our long memory we have never had foreign visitors," Vasily, the chief, a shy man with doelike eyes, tells us. "We don't know why you are interested in us," he says. "But you can travel with us if it pleases you."

They are four families comprising fourteen people: four women in their 70s and Kayta, the daughter of one; three men and their six sons; and 13 dogs. They have no permanent residence, and no memory of ever having had one. They travel year-round with their herd on a route that covers four seasonal pastures and goes from taiga to wet tundra, where north-flowing rivers and streams bleed into the Barents Sea.

The herders' belongings are packed on 84 sleds. They don't ride their reindeer as the Sayka people to the east do, but instead harness the animals to hand-carved wooden sleds. Moving with the seasons and the weather, they live in large skin *chums* (pronounced CHOOMS). "We can't remember a time when we didn't live this way," Vasily says, and invites us into his chum for tea.

We duck through the reindeer-hide flap. Inside, Marie, Vasily's 72-year-old mother, is feeding split birch logs into the woodstove. She scoops snow into a pot to melt for tea. Another pot burbles with chunks of reindeer. *Chai-pi* is not just tea and cookies, but also stew with potatoes, wild berries, and bread.

We sit on skins laid over pine boughs at a low table. Marie thrusts plates of food at us. "You've come a long way through the sky. You must be hungry," she says.

The chum is spacious, 25 feet in diameter. Instead of the sacred pole that stands at the center of the Nenets tepee, the practical Komi set their sheet metal woodstove in the center of the chum and thrust the stovepipe up through the open hole. Marie and Vasily, who is her youngest son, sleep on one side; Piotr, the oldest son, sleeps alone on the other side. Piotr is 52; Vasily, 46.

Through the one window, snow-encrusted conifers sway, and dogs lie curled on the seats of the sleds. It's early April and still cold, with three feet of hard snow on the ground. At 24 miles above the Arctic Circle, almost 67° N, it's unusual to see trees. But birch and spruce are here, thanks to the warming influence of the Gulf Stream. At the same latitude in Greenland or Arctic Canada there would be only rock and ice.

A Komi year is one of frequent movement and harsh winter and summer temperatures, 30 below in the winter, 90 above in the summer, with ferocious bugs. The Komi have reindeer to eat and reindeer from which to make clothes and shelters, with enough animals to sell so they can buy whatever else they need from the village. Winter is spent in the taiga, in the mixed forests; spring and fall are spent on the open tundra near the coast of the sea. In the summer, the Komi move up to a low mountain to get away from the mosquitoes. "Up there the women collect enough berries and mushrooms to last through the year. Then we move south to our winter camp at Golayga. We have to wait until the rivers freeze in order to get across safely." If they do freeze at all. Lately, that has not always been the case.

By the time we finish our tea and reindeer stew, it is dark. Piotr lights the kerosene lamp hanging from a carving he made of a man's elongated face. I had noticed there were no children or young women in camp. "Are any of you married?" I ask. Vasily and Piotr shake their heads: "No women want to live this way, out on the tundra with reindeer anymore," they say. There will be no new generation to carry on the Komi tradition of reindeer herding.

MORNING. Bands of pale light slide down the peeled spruce poles. There are reindeer under me, reindeer around me. I sleep soundly. Someone gets up and starts the fire. The stove wood crackles. Water boils. I sit up. Marie smiles: "Good morning," she says in Komi. She is fixing a breakfast of stewed berries, boiled reindeer meat, and bread.

It's moving day. Outside, thick fog creeps in. Rime ice hangs in trees, on sleds, lacing the net fence, a portable corral in which the reindeer will be herded. Piotr is splitting wood. He looks toward the horizon, a gray blank. "They're hunting for the loose reindeer. They're out there somewhere. But it's hard to find them in this weather."

We pack our things. Then the "village"—the three chums—is dismantled. Stick by stick, the hole in the sky is gone, the center stovepipe laid on the ground. The three boughs at the entrance are stacked on top of a loaded sled. Both men and women unwind the canvas and skin coverings and pull them from the poles.

The loads are diverse: Three sleds carry nothing but chum poles; another sled is laden with a stack of five-foot-long logs. Antlers and spruce boughs are stacked on top of winter clothing. Another sled holds the wood cookstove.

The fog that came in earlier has dropped down to the ground and it is almost impossible to see. Someone hears reindeer coming and yells *"Aleyne!"* ("Reindeer!") The women hurriedly pull on their skin dresses over long underwear and stuff the footpads of their leggings with dried grass for insulation. They grab long poles and rush outside. With their sewn-in hoods tightly framing their faces and fur mittens on their hands, they look like medieval troubadours.

The wait is long. The reindeer were near, but now they are gone, the men tell us. The older women—Vasily's mother and Katya's, also named Marie—and Stas's wife, Rima, lean on their *khorey*, long poles used to prod the reindeer. The fog is a cold room that holds us, now that the chums have been taken down. Hours go by. No one seems concerned that we will be making a late start. A fire is built, a pot of tea is placed over the open flame. A few chunks of reindeer meat are threaded on a stick, roasted, and passed around. The men joke about not having wives. "If we put TVs in the chums," they say, apparently forgetting that they have no electricity, "then we could get women to live with us." Vasily speaks up, "I will bring a television next time we are near a village." Everyone laughs except Katya.

Bright eyed and beautiful, she's 38 and on the tundra to help her mother for half the year. The other half she lives in a town and has seen the devastation caused by runaway capitalism without rule of law and by the epidemic of alcoholism. "We live better here than in the village," she says. "It has always been that way and will always be." When one of the men from her chum suggests building a casino in Pesha, Katya turns to him angrily: "If you play in the casino, I will kick you out of the chum."

SOMEWHERE OUT ON THE TUNDRA, Rima's bearded son, Nikolai, is looking for the reindeer. There are two herds: the gelded reindeer used for pulling sleds and the big herd of breeding males and pregnant females that are about to begin calving.

Reindeer hunting and husbandry have always been the traditional way of life for indigenous people of Arctic Russia because tundra can support no other crop or herding animal. Biologists have put the domesticated reindeer population at 1,357,700 animals; there are reportedly, 1,246,000 wild reindeer, whose wide feet are superbly suited for the frozen habitat of Russia's far north.

The word "tundra" itself comes from the Finnish, *tunturi,* meaning "completely treeless heights." Tundra is low in biodiversity and supports only a handful of mammals but is home to more than a thousand species of flora and thousands of birds and bugs.

Permafrost lies under 25 percent of the total land area of the Northern Hemisphere, its presence depending on the balance between geothermal heat from the Earth's interior and the average annual air temperature. Russia's far north has more permafrost than any other nation, but as the world warms and wildfires spread across the Arctic, subterranean ice is imperiled, and the greenhouse gases it sequesters are being released.

Tundra has peaty gley soil carrying continuous mats of mosses with patches of lichen, dwarf willows, sedges, cotton grasses, berries, and such Arctic flowers as saxifrage and dryas, but it is the lichen "starch," full of amino acids and vitamins, that nourishes and sustains reindeer throughout the winter. In turn, the tundra-adapted animals sustain the lives of these northern people: Komi,

Nenets, Eveny, Sakha, and Chukchi, among others. But as the Earth warms and the permafrost melts, even this marginal ecosystem will become toxic, a place of methane emissions, radioactive leaks, breaking pipelines, expanding lakes, reduced vegetation, and ultimately another heat sink. "A place only a mosquito would like," one of the herders says.

AFTERNOON. Nikolai floats through a sea of reindeer. We cannot see his sled over the moving animal mass, only his shoulders and head. They swarm over the hill panting, and in the fog their breath spills out in white plumes. They approach the portable corral warily, flanked on all sides by the old women and their sons. Finally, they flood in, one black dog barking at their heels. Packed tightly, they're a boiling mass. The reindeer push at the fence, and the women lean in, holding the top edge of the net high. Antlers clack, brown noses stick up out of the swirl. One huge male rises from the crowd, pawing at the animal in front of him, and accidentally paws Katya in the face. Tears come but she waves me away, indicating that she's all right.

Soft ropes fly: A reindeer is caught and struggles backward. Another one is roped, and another, until the first six are pulled from the moving mass. Katya, Piotr, both Maries, Stas, Rima, and Red Beard, one of the herders, hold the herd in. Antlered heads rise up and drop. Strips of velvet hang from broken tines. A young herder named Alexander, Katya's brother, pulls out six more. The rime ice on the net corral jangles.

The men know each animal, which ones worked last and need a rest, which ones are fresh, which ones are young and need to be put next to an experienced animal. Harnessing is slow, but no one ever looks hurried. Each member of the group drives a lead sled that pulls a caravan of packed sleds. Fourteen drivers require at least fifteen reindeer each: three or four to pull the first sled, then two more between each of the six or seven freight sleds.

It's late afternoon by the time we take off, but because it's April and we are above the Arctic Circle, there will be light well into the night. Women lead the way. Katya's mother heads out first, followed by Marie and Rima. They are strong drivers and hold their khoreys parallel to the ground like lances. They know the route, as do the older reindeer. The men's job is to decide where camp is set up and which direction to face the chum door.

The small sleds glide across the ice-covered ponds and bump over purple hummocks. Katya and I are fourth in the caravan. We share the single seat, really only big enough for one. Her reindeer run at first, and we bounce hard, then they slow to a trot. There's only a single rein attached to the bridle of the lead deer, with little real control. The dogs are tied to the sides of the sleds and run along happily. Puppies get to ride.

The Eveny people to the east say that when their shamans wear antler headdresses, they become reindeer that can fly. Reindeer, thought to be special because of the hypnotic look in their eyes, are consecrated as sacred and called *kujjai*. They are protectors who keep their owner from harm, even dying for him or her if necessary. As our sled, pulled by four reindeer, bumps and slides past snow-flocked spruce and open meadows, I can't help thinking about the cascading disasters Russia (and every Arctic nation) is facing here in the far north, with its acid rain-laced tundra ponds and radioactive lichen, yet it feels as though we're in a past century, before these anthropogenic calamities occurred. Traveling in a wintry shroud of snow and fog, we are clothed in hide and furs, and the clacking of reindeer feet made by a tendon rubbing across a bone in the foot sounds out a primordial rhythm.

After a long traverse, we head up into the forest. The track through the trees narrows, and we have to duck under overhanging branches as we come to the top of a hill. Below is the frozen Snopa River. To get down the steep ravine and cross the river ice, then climb back up the other side with 87 sleds will take hard work and time.

Katya and I wait our turn. We have been traveling for hours, sharing our narrow seat. While we stretch our legs, the reindeer lie down and eat snow. Katya strokes the lead reindeer's forehead. Flaps of velvet hang from one tine. "He's my favorite," she says. "They always get this one out of the herd for me."

The Saami reindeer herders to the west, who are linked to the Komi-Zyrian people by their Finno-Ugric language, have a legend about a "wild woman" who was tired of being a human and so became a reindeer, though she still had human thoughts. As a human, she had been childless, but once changed, she gave birth to a reindeer. "That's me," the unmarried Katya says, smiling. "I would take this old reindeer for a husband any day."

Fyodor, a Nenets and the youngest of the group, glides up beside us, jumps off his sled with a flourish, cocks his fur hat to one side, and runs down the hill to help the others. Each sled is eased down the embankment and hoisted over boulders. Down they go, fishtailing across the frozen river and bumping up the other side, three men pushing from behind.

*Rangifer tarandus* are actually caribou that are called reindeer when tamed. Domestication is only partial. They are simply separated from the wild herd and put into harness, receiving no special feed or shelter to lure them into the sanctuary of the human world. You can look into their eyes and not know what they are feeling, whether they are frightened, bored, angry, or just don't care. They are not communicators, as dogs and horses are. They only tolerate the harness, preferring, most likely, their other life on the range.

Caribou biologists say that domestic and wild herds are "ecological antagonists." They compete for feed on the lichen pastures and are sometimes carriers of the deadly anthrax disease. The life span of wild reindeer is only half that of tamed ones, who live to be 14 or 15 years old.

When it's our turn to cross the river, Fyodor takes over and Katya and I run alongside. "He's showing off," Katya whispers to me with a smile. Traveling again, thick trees give way to a series of huge meadows, at the far end of which a new camp comes into being.

Despite the lateness of our arrival, the camp is a hive of activity. It's ten in the evening by the time the first chum goes up, the two center poles forming an A-frame and the other poles carefully balanced against them. Four wooden planks, the portable floor, are put down on either side of the woodstove, and cut pine boughs are laid around the perimeter with reindeer skins. Fires are started in the sheet metal stoves so that by the time the chum is enclosed, the stove is hot enough to heat water for tea and cook the evening meal. Across the skeletal poles, the men unwind the hide cover as big as a sail. It takes 25 reindeer skins to cover one chum.

Logs from the forest are ferried in on reindeer sleds, and firewood is cut. At our chum, the last to go up, Marie bosses the men as they push the reindeer skins to the top of the chum. "Not too high, down on that side . . . no . . . no, OK, that's right, higher, it's going to be bad weather tonight," she yells, as if they'd never done it before. Unruffled, her sons do her bidding.

The back wall of the tent with the sewn-in window is carefully adjusted to let light in on the low kitchen table. An oilcloth is smoothed out, and on it Marie lays dishes, silverware, tea, cookies, and candies. Chai is poured. A candle is lit. The movable world of the Komi people is again in place.

One o'clock in the morning, and outside the men, having worked late, rest on the huge pile of firewood they've just cut and tell stories: "Two years ago in December and January, in just this place, three wolves came into the herd and ate some of our reindeer," Vasily tells me. "There was very little snow, so the wolves could get away from us. But when they came back, the snow was deeper and we were able to shoot them. We didn't eat the meat, but we used the skins."

That same year a bear attacked three harnessed reindeer near their chum. "We were eating when the dogs began barking," they told me. By the time they ran outside to kill the bear, the reindeer were already dead. The bear stayed around camp all night and found their cache of meat. "It ate that, too. It ate everything." They look at the woods nearby. "Maybe we will have a bear come into camp tonight," they say, smiling. "They are smarter than we are, so watch out!"

The evening brings no wolves, no bears, only sun touching down on reindeer hides. Snow falls. Marie ties a red wool scarf over her head, shuffles out away from the camp activity, scoops clean snow into three buckets, and hauls them into the chum to melt for tea. The dogs are fed and curl up under the sleds for shelter.

The sky darkens and the spring air is cold. I follow Vasily and Piotr through the flap of the chum. It's warm inside. Vasily looks boyish, with brown bangs and big, soft eyes. He tells how when it was time for him to go to boarding school he hid when the helicopter came for him. "We were on the tundra, but there were some trees nearby. I ran into the forest and dug a cave in the snow, but the pilots found me and dragged me away."

Like children all over the Arctic, Vasily found this enforced separation from camp life and de-acculturation traumatic. He and the other children spoke only Komi when they arrived at school, and it took an extra year of classes just to learn Russian. They said they couldn't digest the food. "Especially porridge," Vasily says. "Before then, we had only eaten reindeer meat, fish, and berries." As he talks, Marie makes his bed with loving care. She piles up skins, positions a large pillow against the chum wall, and lays out his sheepskin bag at a right angle to her bed so that his head is almost in her lap. At 46 he's still sleeping with his mother.

Piotr turns on the radio. It's cheap and battery powered and blasts only static. Then voices do come on, something about Moscow's weather followed by polka music. He's the restless one of the two brothers and fends for himself. His spartan bed is on the other side of the chum. He asks if I know Madonna and looks disappointed when I shake my head no. Mike Tyson?" No. "Tina Turner?" Again I shake my head, laughing. He says: "I know more about her than you do: I know she's 58 but looks 28." Vasily gives his brother a dirty look: "They can have plastic put on their face, but they can't fix their health," he says. "You shouldn't be interested in these people. They think we are poor. But we use everything. No, we are rich. We are healthy living this way."

Piotr says nothing. He lights a cigarette and looks up at the smoke hole. He turns to me once more: "Will you vote for Hillary Clinton?" I'm not sure, I say, though when I ask why he wants to know, he can't give an answer. Piotr is wrestling with ideas about personal and political freedom. Russia's contemporary upheavals, from tyrannical dictatorships to totalitarianism to perestroika to the iron rule of Putin, haven't garnered him the freedom to travel, to vote, to voice his opinion, or to make enough money to go to art school. A few years earlier he left the tundra for a job in

Arkhangel'sk, but he didn't like city life so he came home. "They didn't pay me enough, and I had no free time. I hated it. Not enough money to rent an apartment or have a girlfriend; no time to do carvings."

From the other side of the chum Marie's snoring gets louder. She shifts and groans. Earlier, I heard a snow bunting's bell-like song. Now, burning birch logs in the woodstove click and fizz. Tundra sounds seem to come one by one, the grunt of a reindeer, people sleeping, a radio going dead.

There's a bump on the side of the chum and the tent flap flies open: Young Fyodor slides in on his knees. He sits on a low stool facing me, eyes glistening from vodka as he cocks his fur hat. He wants to talk. He wants to tell us about his life, and before anyone can object, he begins:

"My mother is Nenets, my father is Komi. I grew up in Oma. It's a Nenetsy village. That's where my mother is from. But when she got married, we moved to Pesha to work with reindeer. My mother stays in the village and shows movies once a week at the community hall. I grew up on the tundra. That's why I'm here now. I served two years in the elite tank unit in Moscow. I had been out here with the reindeer, but the guys from Pesha said I had to go to the army, and they took me to the waiting helicopter.

"When I got to Moscow, the army psychiatrist wanted to know what went on in the mind of a nomad. I told him: 'Reindeer.' Two years later, the same guys from Pesha came and brought me back. So here I am!" he exclaims with a comical smile, jumps up, knocking the stool over, and abruptly leaves.

Piotr fiddles with his shortwave radio again. This time, news about Chechnya comes on. He holds it closer to his ear: "We are interested in what goes on there," he says as the voices fade. In frustration, he shoves the radio under a jacket, and still fully clothed, pulls his single reindeer skin up to his chin. "I sleep with my boots on," he tells me. "It's from living outside with the reindeer all my life. I would prefer to sleep outdoors all the time."

Six miles away, Katya's two brothers are camped out with the main herd of 2,500 reindeer. Despite the snow, they have no tent. "It would make us lazy and sleepy, and then the wolves and bears would come and eat the calves," one of them tells me. Guard duty lasts a week, then two other herders will take over. "We were 12 years old when we began to stay out with the reindeer," he tells me. "It's hard sometimes, but we like it. Our reindeer clothes keep us warm."

Next week Vasily will live out on the range with the expectant reindeer, but for now, he's dreamy eyed and quiet. He sits back against the tepee poles with his mother's sleeping head against his thigh and speaks slowly, with the quiet reserve of a much older man: "We like it here, living in a chum, because these poles we have carved from the forest, and these walls we have sewn together from our reindeer. That is the meaning of home."

Clouds wheel by the smoke hole. There has been no sun during the day, no moon or stars visible at night. Directly north is the shallow Barents Sea, whose influx of warm Gulf Stream water from the Atlantic moderates the ecosystem. In the summer, beaches are blanketed by colonies of nesting seabirds: arctic terns, ivory gulls, Sabine's gulls, little auks, and pink-footed and barnacle geese. The huge bearded seals, narwhals, and walrus breed and rest along the shore. Cod is abundant. But the sea and all its life are polluted.

To the west is the Kola Peninsula, which has the highest density of nuclear reactors in the world. A former naval storage facility for spent nuclear fuel and radioactive waste, its storage tanks are leaking, and radioactive contamination is migrating into groundwater and the surrounding atmosphere. (As if acknowledging its environmental problems, the Russian government has agreed recently to develop wind farms there.)

The richly productive Barents Sea is a dumping ground, where 7,000 tons of solid radioactive waste and 56,500 cubic feet of liquid waste have been deposited. Eighteen nuclear submarines were abandoned there, giving off cesium-137, cobalt-60, strontium, and iodine worth 312,500 curies. The Chernobyl disaster gave off 50,000 curies; 1 curie is enough to kill a human.

Rising sea-surface temperatures in the Barents Sea, northeast of Scandinavia, are the prime cause of the retreating winter ice edge over the past 26 years, according to research by Jennifer Francis, associate research professor at the Rutgers Institute of Marine and Coastal Sciences. "The recent decreases in winter ice cover are clear evidence that Arctic pack ice will continue on its trajectory of rapid decline," Francis concludes.

To know nothing of these things, is that happiness? Perhaps to know and be powerless is worse. Piotr blows out the kerosene lamp and makes his way back to bed by match light. The tent shudders; a frigid wind blows. Above and below transboundary pollution, beauty makes its mark on the ear, the retina, and the heart. Here we are living in an elegant reindeer universe: Chum coverings, clothes, sleeping bags, food, thread, glue, and transportation come from this one animal. Unlike hunters on the ice, who search for prey they cannot see, whose whereabouts are undependable, the herders take their food supply (reindeer) with them.

When everyone is asleep, I stick my head out the flap. Snowflakes tap my cheek. Taiga is my headrest and tundra my bedcovers. Smoke curls up from every chum, and the dogs sleep, noses tucked under tails.

MORNING. I ski alone across the open meadow, down to the Snopa River, with no idea of where it might lead. A hard crust formed on the snow during the night. Russia is a country of beauty and abuse. Under one ski are leaf-plastered shelves of ice at the river's edge; under the other is coercion: Stalin's terror, the tyrannies of Communism, and the surge of capitalism without a fully realized democracy.

Siberia was the Russians' frontier before it became their dumping ground. First contact with indigenous people between the Dvina River and the Ural Mountains had occurred by the early 1400s; they were "conquered" by 1456, and by 1620 the annexation of western Siberia was deemed complete.

I asked the Komi how they were treated during the Stalin years and later by the Communists in the 1950s. "They forgot about us," Vasily said. "We weren't collectivized like the Indiga Nenets were, their reindeer taken, herders and their families removed from the tundra, the men forced into all-male brigades. They left our families alone." Was that because the Komi look like Russians and are therefore racially "acceptable," or were their numbers too small to be bothered with? Vasily shrugs.

To the east, the Nenets on the Yamal Peninsula rose up against Soviet authority in the 1930s. They were spirited and independent and had no need for Russian things. Soviet Army men came onto the tundra of the Yamal and demanded to be given reindeer and sleds. In some cases thousands of animals were taken, leaving behind only a herd of a hundred. Shamans were incarcerated or killed. One powerful shaman who was inadvertently left behind called for a "holy war" against the Russians and a *mandalada,* an assemblage, was formed. The leaders made demands: that the stolen reindeer be returned, imprisoned shamans be released, their children not go to Russian schools, trading posts be banned, and

Soviet laws be denounced. But time and again, the Russians over-powered the Nenets rebellion. Eventually, the tundra Nenets leaders were arrested.

The Cold War of the 1950s and 1960s left no one untouched, but the hand of radioactivity is hard to perceive. Just to the north of this camp the radioactivity blew east in the prevailing winds all the way across the Siberian tundra, only to be sucked up by lichen, which take their nutrients from air not earth. These, in turn, are eaten by reindeer, and the reindeer are consumed by reindeer herders and townspeople alike.

I ski and ski. Under the snow and the active layer of earth that freezes and thaws seasonally are bodies of ice, lenses and veins of it inside rocks, under hummocks and mountains, even under ponds. What floats on top is the thinnest veneer of matted green, but there's a spangled ice-rich garden beneath that has grown and changed shape unseen. Farther east, toward Siberia, huge ice wedges have joined into what Vladimir Kotlyakov, a geographer who has focused on environmental science and glaciology, described as "an almost continuous ice massif" that in some places is a mile thick.

Now Russia's permafrost is feeling less permanent. It is beginning to melt, outgassing millions of tons of methane a year. German climatologist John Schellnhuber from the Potsdam Institute for Climate Impact Research in Germany told me that once it gets going, "the air will smell like rotten eggs and Earth's 'fever' will come on ten times as fast." Methane has 25 times more effect as a greenhouse gas in the atmosphere than carbon dioxide. Though summer-thaw ponds and bogs have always released methane, more of it was taken up and held by the permafrost itself.

A thawing landscape is a wreck of a place: broken ground, heaving mud, and ever expanding lakes that paradoxically dry up, leaving no water or vegetation at all. If warming continues rapidly, tundra fires will break out in summer heat, burning underground and thawing even more permafrost. The perfect union between caribou, herder, taiga, and tundra will be lost.

I return to camp to find a production line of tepee poles being peeled and cut to size. Drawknives are pulled down long pieces of spruce. Perhaps hope for the future resides right here, where people are still making the things they need to live. During a lull in conversation I ask if there have been any noticeable changes in the climate, in bird migrations, storms, temperatures, or ice conditions. They say they don't know.

Finally, Vasily says, "The climate is not changing so fast for us yet. Not like it is for others. I have heard about it on the radio. Some things here are even better. In 1975 and 1976, there was a hard crust on the snow here until late May, and it was hard for the reindeer to paw through for something to eat. Now it's easy," he explains.

Nikolai pipes up: "Some river crossings are becoming a problem. I heard that the Nenets on the Yamal Peninsula are having trouble crossing the Ob' River to get to their winter range because the river is still open in November. We have that trouble too, crossing the Pesha River. Small ones, we can make a bridge, but on the Pesha we have to float across."

Autumn is the most difficult time. Vasily says, "Bad weather always makes it hard to cross the tundra." Are the storms worse? I ask. They're not sure. "Sometimes it is ice, sometimes it is open water, and there's fog and rain. Later, the reindeers' antlers get covered with ice, so there is no water transpiration into their bodies. It's becoming a problem. Now it's happening every day at that time of year, whereas before, it was not so often."

"In the last years we had fewer mosquitoes, but more biting flies," Piotr says. "The aleyne, the reindeer, are very bothered by them. They go through the skin and lay eggs. It hurts them, and also it ruins the skins. And yes, spring is beginning a little sooner almost every year."

They quickly change the subject to a reindeer sled race to be held in Pesha village next March in which they'll take part. "We'll need three or four of the best animals for each sled. The leader has to run straight. No jumping. If he jumps, he's disqualified. We want to win because first place is a snowmobile and second place is a generator." The irony of using a reindeer sled to win a snowmobile seems to escape them.

KATYA HAS INVITED ME to her chum for supper. As she cooks, she teaches me Komi words: *skurr* for rain, *patch* for woodstove, *purga* for storm, *vur* for forest, *shor* for creek, *yu* for river, and *tailus budma* for waxing moon.

"I come here to help my mother for half the year," she says. "My sister comes when I leave, to take over." They live in the middle and largest of the three chums. Katya explains that two separate families have lived for more than 80 years in this one chum divided by only the patch, the woodstove in the center. "At one time there were 20 of us living in here, and 16 of them were children. Now the numbers are smaller." She and her two brothers live with their mother on the left side of the chum; five men live on the other side. "What happens if someone doesn't get along?" I ask. She looks at me quizzically, as if I didn't understand that they live rich lives of intimacy and cooperation, that any other kind of behavior would be uncivilized. "That never happens here. If there's a problem, we talk it out among ourselves or send them to town for a few days."

Vasily and Katya's younger brother leave to take the place of the other two men who have been tending the big herd for two weeks. They come into the chum, dirty and hungry but in a fine mood. An old dog follows them and lies by the fire. "It's OK. He's a pensioner," they say, smiling. I ask how it was out there in bad weather. "We don't have a tent. We just live outside on the snow. It's OK, but we were down to our last cigarette," Katya's older brother laughs. "We broke the last one in half and shared it. Then Vasily showed up and gave us some, and so we survived!"

Rain stops. The warm air and melting snow produce fog. We're in a wide opening. Behind us toward the river are trees. The days of frost are over, and the time for skiing is quickly passing. But without skins or snowshoes, we sink to our thighs in three feet of snow. The nights are white, and the ice-covered sea to the north is becoming blue. Ponds leak. The season is neither this nor that, neither winter nor spring, but something between, where the solid becomes liquid, a tawny tundra broth that will later be boiled for tea.

In the 13th century the Komi were ruled by a prince who lorded over a vast area of arable land called Perm. The Komi were farmers then, but when small groups of them moved north, they learned to herd reindeer from their neighbors, the Nenets, with whom they have frequently intermarried. "I'm a Nenets," Stas tells me, "but I'm married to Rima, who is Komi."

At one time the Komi were drafted by the Cossacks to fight the Khanti, Mansi, and Nenets. Later the old divisions vanished, and all nomadic reindeer herders were subjugated by the Russians and forced to pay taxes in furs. These "small people," as the Russians referred to them, were vanquished by smallpox and robbed of their reindeer wealth.

Vasily says that a week ago he visited the Nenets camp near Indiga. "They are staying 20 kilometers [12 miles] from the sea, but there is only one tent in camp because the men live without

117

any women. They stay on the tundra for a month, then go back to the village, and other men take their places. The reindeer don't know them. It must be difficult. They live in 'brigades.' They are no longer *yumdai*, 'always moving.' "

Vasily takes pride in the fact that he and his group still live nomadically. "We do not live against the tundra, against the reindeer. We move with it," he says. Over and over he intimates how quickly the "open soul" of the herder can be corrupted.

"We don't make rules for each other. We know each reindeer, each person in camp. If the reindeer goes straight when he is harnessed, we see that; if he runs away, we see that in the person handling them, too."

Life on the tundra is not romantic. Vasily says, "A hard time of year is right now, in the spring, when we have to move so often. But it is harder in June when we are moving every day to get away from the mosquitoes. The reindeer run so fast, we have to leave the loose ones far behind and go back for them in the morning."

He muses silently for a while, then says softly, "You have to know what the reindeer are thinking to live with them the right way. Autumn time, when the mushrooms appear, they run to them. There are all kinds of mushrooms. We like them too; we salt them and save them for later."

In northeastern Siberia, Chukchi reindeer herders declare that "the owner of the world is Earth." They eat red mushrooms that are hallucinogenic, believing they represent "a separate tribe." According to Waldemar Bogoras, a noted anthropologist, the mushroom spirits are thought to be very strong. When the mushrooms come out of the Earth, "they can lift a large tree-trunk on their heads and shatter rocks to pieces." They lead humans on intricate paths in the world, enabling them to see in those places what is false and what is real.

"The Komi don't eat that kind of food," Vasily tells me, nor does he remember having shamans. "A long time ago, there were forced baptisms, and it was then that we became Russian Orthodox Christians and are still believers today."

The Nenets on the Yamal Peninsula say that a mark in the shape of a drum appears on the body of a child who will become a shaman. A "spirit master" will come in a dream to that child. Later, these children go through the usual period of isolation and initiation, and after, wear a special coat hung with metallic pendants, images of the sun and moon, bear teeth, raven wings, talons, and claws much like the belt the Komi wear, though they say that their ornaments are merely decorative.

The long prodding stick the Komi carry could have been the sacred staff carried by a Nenets shaman, with the carved head of a spirit protector on top and used alternately as a divining rod to find water. These powerful shamans would select a site for a séance in a place representing the conjunction of the "three worlds:" river, tree, and the cosmos, or tundra. Nenets believe that the most important temple is not the one people go to on Sundays but an island or any remote and inaccessible place. The least visited is the most sacred of all. A Nenets elder, Avvo Vanuito, once explained that "the major gods live at the end of the Earth."

RAIN IN THE NIGHT. Then the sky clears and we can see a slim moon and Venus. To the northeast the horizon is blue. It looks like the sea, but it isn't. To the south a hard-blowing wind has erased ground and sky. The visible simply ends as if expunged; three leafless birch trees are the gateway to the barren tundra beyond.

By morning the white world of flocked trees around us has vanished. I wash my face using water from a small iron pot hung by a chain. From its spout, water pours into my hands. Outside, the trees are black arrows shooting up from snow-covered ground, pointing toward a new season.

The net corral no longer holds ice. When the reindeer are brought in, the sun sends spears through clouds, but once again a dense fog descends as the Earth warms after a long Russian winter.

Alexander hacks a deer head into quarters and throws the bits to his dogs. Sixteen reindeer are taken out of the herd and harnessed. Some are young and wild. When Arthun, one of the youngest herders, who came from a town to the north, harnesses three of them, they take off in the wrong direction. The older men stand and watch. It is a long time before he can turn them again.

In the middle chum Katya folds clothes and helps her mother pack the kitchen: Porcelain cups go into a padded wooden box with a lid, plus the plates, spoons, forks. The reindeer skins are stacked and loaded on a sled, then the walls of the chum come down and we stand exposed to the snowy world. The spruce poles are dragged away, the skins folded, the kitchen table is gone, and the ashes from the woodstove are dumped. Sleds are repacked. We are traveling again, one in a series of moves that will take us to spring camp.

The day is cold, and we bundle up to travel. To live nomadically in western Siberia doesn't mean one is homeless. Quite the opposite. Home is wider than four walls. Home is the wall and roof and floor of each season. White, green, and brown. It is taiga and tundra, mountain and river, lichen, moss, berry, reindeer, and bog.

Fourteen long caravans clatter down from high ground and patchy forests. One of the dogs, Buryan, named for a Russian snowmobile, not only herds reindeer but also, when tied alongside, helps to pull the sled. It's rough going at first, but ahead is a long straightaway. Suddenly, we glide, the harnessed reindeer trotting with the dogs and loose reindeer running alongside. Taiga and wintry weather will soon be behind us, open tundra is ahead.

We come to the snow road that leads to the village of Snopa. The reindeer clamber up the road's steep side, and instantly the ride becomes smooth and fast. For a while things are split in half: winter and snow on one side, summer and sun on the other. The trees thin out. Then we leave them for a cool vastness, an ice-flattened world of brown hummocks encircled by tangled vessels of slush, water, and ice.

Here and there dwarf willows poke up, so short that only the topmost leaves show. Mosses rule. They are closed communities that stabilize the soil temperature, hold in moisture, and discourage other plants from taking over. But lichens are life giving. They produce a sweet "starch" made of polysaccharides, amino acids, and vitamins that is ingested by reindeer and is used by birds for nests. When we stop for a rest, I kneel down, scrape away snow, and press my hand in. The tundra is a sponge, but a fragile one. Lichens, if undisturbed, grow only two-tenths of an inch per year, and if reindeer overgraze it, the tundra mix can take six years to regrow.

As the day warms, fog returns. With little visibility the caravans become separated, and we see the tail end of the last sled vanish. Snow blows onto the tracks. Andrei, my translator and guide, takes a compass reading just in case we get left behind. Flat land, flat all the way to the cod-rich, polluted Barents Sea. We pass the tilted dunes of tundra, thick with tiny yellow berries and small orange flowers. Another bend and another, and finally we see the other sleds in the distance. Half an hour later, we're at camp.

Already the first chum is going up. Even this close to the Arctic Ocean there are trees in view because of the warming influence of the Gulf Stream. This is spring camp, and the chums will stay here until June. Sun shines through the fog, but the wind is frigid. "Maybe the wind has come to take the fog away," Marie says. She sinks all the way to her knees as she collects snow to melt for water. Inside the newly assembled chum, birch-bark peelings, gathered a day earlier, are used to light the fire. She pulls out a loaf of stale bread, picks a reindeer hair from the top of the cranberry jam, and cuts reindeer meat on a hand-carved cutting board.

We're camped near a forest island called Kol'-Ostrov. The night before I dreamed that it exhaled gyrfalcons, brown bears, and swans. Now Andrei and I ski to it. He is on wide Komi skis with old leather bindings, pushing himself along with a pole because, he says, he wants to live the Komi way. The Komi make their own glue from boiled reindeer antlers. Those from the males are best, the herders say, and with it, they glue reindeer hides to the bottoms of the skis to keep them from sliding backward. "The best skin to use is from the river otter. We glue it to the bottom when we're hunting. It's very soft and makes no noise," Vasily has told me. In the spring the width of the ski is more important than the swiftness of the glide. The Komi skis function more as snow-shoes, as a way to float on softening snow.

We stop for a snack, and for a moment, I wonder where in the world I am. We've come a long distance by plane, train, helicopter, snowmobile, and reindeer sled, yet it all feels familiar, like home. Not just the landscape but the way of living: shoveling snow, cutting wood, cooking on a woodstove, melting ice and snow for water. Distance and language are not the only things that matter. It is how we live wherever we are, close to the ground, shitting in the snow, sleeping on pine boughs in a circle of humans and animals.

Andrei, trained as a biologist and an avid naturalist, looks for bear tracks, wolf tracks, but sees none. Up high there are falcon nests, but because it is early in the season they are still empty. We cross a melting cranberry bog, dropping through one layer of ice and water to a second, firmer ice floor. In the distance, shifting lake ice booms. A swan flies into the Kol'-Ostrov. "The Earth is waking up. You can hear it," Andrei says with a smile. On the horizon another small lake appears. The Komi call it Happiness, because it is home to so many birds. The Komi love birds and favor the ptarmigan that come and stay in camp; they say that when the geese arrive, the reindeer will start calving.

EVENING. Blue sky and open country. Fires in every stove. Reindeer grazing in the distance. This is the camp where the women will stay until calving is over in late May. Mu Mu, one of the small black dogs with soft black hair and tiny stand-up ears, and another dog sit together on the seat of a sled and howl, as if to say finally, it is spring.

Katya sweeps the floor of the chum with three raven wings. I ask her if the feathers are a sacred amulet. She laughs at the suggestion. "No, they're just for cleaning." She's a wide-eyed beauty, vigorous, affectionate, and innocent. Outside, on the tundra, in her hooded *malitsa* she's a medieval nun, but in the chum, wearing black tights and a turtleneck, she's modern, efficient, quick-witted.

In June she'll return to the small town and her older sister will come to camp and take her place. "Every year I say, 'This will be the last,' but then I come back. They say about people like me that I have tundra fever. Well, maybe I do. I love it here too much to stop coming," she says.

We dig trenches to keep meltwater from seeping onto the floor of the chum and lay firewood in front of the stove. A sharp wind comes up, then stops. Katya takes off her malitsa, hat, and mittens. She washes her hair. With it dripping, she says: "Skurr!" Rain. We run outside to snatch our still wet clothes from the line and hang them over the woodstove.

Katya's face alternately registers excitement, sadness, and calm. "I feel very good now, inside of myself," she says. "I had bad experiences with men and also because of racism, because I am not Russian. When I went to college, people made fun of me because I was Komi. And the men, well, there was always too much vodka. Now I am a woman not looking for a man, a woman who lives in two worlds. This is best for me. Yes, maybe this is the only way."

Late at night Vasily returns from Pesha slightly drunk. As if ashamed of him, his mother lowers a cotton curtain over their two beds. In the dark, Piotr lies on top of reindeer skins, smoking. He says the Nenets living nearby are having problems keeping their language alive. "They have given up living all the time with their reindeer. We Komi are still *vetziny,* nomads. I'm proud that I speak Komi and wear a malitsa. I prefer this way of living, always moving with the animals and our families. I lived in town once and worked. I know what town is. Living with the reindeer, making everything we need, and requiring very little else. That means we are free."

He sits up and stubs out his cigarette. The night's darkness breaks into something darker. A line of pine and leafless birch trees follows the twists of a river flowing north. Fresh air swoops down through the smoke hole. Piotr says, "If I had a million dollars, I wouldn't buy a house, or a car, or get a wife. I'd travel." When I ask where, he says, "To the places in the world where there are reindeer, to Lapland (Sápmi), Mongolia, and Chukotka." In other words, he would never venture far from home. "But if I want to travel, I must begin getting a passport now because it takes years to get one."

In the morning, Piotr goes ice fishing, and the others gather skis and winter clothes and head toward the village of Pesha, where the gear will be left until late autumn. "We don't like Pesha village because they use coal. It smells bad, and the reindeer can't eat the grass because of the coal dust," they tell me. "When it's time to sell our reindeer, we take the meat to Pesha first, then it's flown to Nary'an-Mar, a city northeast of here. We get $200 for a whole reindeer, and it costs only $250 to fly by helicopter to Arkhangel'sk. But we need permits to travel anywhere. If only reindeer could fly."

Marie is sewing a new heel onto her son's leggings with a needle and caribou-sinew thread. I inspect the stitches. "We make most of our clothing in the winter," she tells me. "Summers, we just make repairs." Mu Mu, my favorite of the 13 dogs in camp, lies in the flap door, shielding himself from the rain. The women gather to have tea together while the men are away. Marie has put on her best: gold earrings and a paisley head scarf. We eat cookies, sliced apples, oranges, and lemons brought from Moscow.

"During World War II," Marie says, "when the German planes came to bomb us, we put white blankets on our tamed reindeer so they wouldn't see them. It worked! The planes came right over us and flew away. We know where the British plane that crashed lies in the mountains. There are still things in it. Guns and bodies. We don't touch it."

She rummages around in one of her hide boxes and pulls out another box of cookies. "We set the door of the chum in different directions, depending on the season," she says. "We call autumn dirty wind because of the fog. In spring the bogs come on with cranberries. We pick them before we go to the mountains. In midsummer, up high, we pick other berries and also herbs that we make into medicines because even now, they are healthier than tablets."

*"Burrasi,"* Marie calls out. "Good morning. Get up quickly. Good weather!" she shouts. Today Andrew, Andrei, Gordon, and I will leave the women's camp and go with the men to the main herd of reindeer that will soon begin calving. Only one man—Rima's husband, Stas—will stay behind with the five women. The puppies are playing, the men are harnessing reindeer, and the women are busy packing rucksacks with extra clothes and food, freshly baked rolls, salmonberry jam, butter, and tea. Then it's time to leave.

I have mixed emotions—wanting to stay here and longing to see the newborn caribou calves. I look for Piotr. He's fixing an ax. I ask him if he is happy living here. "I would like to see more," he says. "It's not enough for me to see just this, to know just this. I can do many things. I can care for reindeer, run an excavator, paint and make carvings, but I want more. I would like to go to art school; I'd carve and paint and draw. Then I could develop a handicraft business and sell my things. I need more space for my life to grow wide."

Andrei is calling to me to get on the sled. In the confusion I search for Katya. She has put on her malitsa so we can have our photograph taken together. Afterward, we stand face-to-face, holding each other. Simultaneously, though in different languages, we both say, "I will never forget you."

Andrei and I climb onto a sled pulled by Red Beard. He's a grumpy man and handles his animals crudely. As soon as we're seated, the reindeer lurch ahead. A piece of wood supporting the runner breaks. I yell at Red Beard. He doesn't seem to care. Andrei shrugs. The sled holds.

When I turn for a last look, I realize that Piotr is also staying behind. He'd wrapped the carving that I'd admired and put it on the sled. How could I have known it was a farewell present? I yell,

"Piotr! *Spasibo!*"—thank you. But he has already turned his back to us and is cutting wood.

The way is not smooth. Only the reindeer and these small sleds could survive in the tundra. We tip and tilt over hummocks, crash down onto ice, slide, fishtail, slam down, dip and drop into melt-water moats, then haul up onto rust-colored humps of lichen and moss, and jolt down again into water. Andrei and I, sharing a seat, grab the sides of the sled to keep from falling.

"How do you like 'tundra-thumping?' " Andrew, the film-maker, asks as his sled bumps by. Being from Barrow, Alaska, he's used to this kind of ride. I raise a thumb. "Bitchin'," I say. He smiles, but when the young reindeer hitched to his sled take off in the wrong direction, veering so fast he becomes airborne, his smile disappears.

A white-tailed eagle flies up, but still we see no geese. When I ask if the reindeer have begun calving yet, Red Beard shakes his head. We follow the river south to another river, where we are going to have to make a dangerous crossing. The sled stops on the edge of the cliff. I peer over the side. It's worse than I imagined: The current is fast and the water is deep, with huge shards of broken ice crashing into each other and stacking up on shore.

I ask the men how we will cross with our animals and heavily laden sleds. They laugh nervously. "We put pieces of wood and logs under the sled runners, so they float like small rafts. The reindeer swim and help pull the sleds across. Their feet are wide, made for tundra walking and good for swimming too, but you have to be careful because the water comes pretty high." "Will there be problems?" I ask. "Sometimes a sled runner gets caught under a piece of ice or the harness gets snagged, and then there's trouble."

"And then?" I ask.

They shrug. "Do you swim?" I ask.

"*Nyet!*" Laughter.

Andrew changes places with Andrei to ride with me. He wants to film us crossing the wild river with the reindeer and whatever else comes next—floating, swimming, drowning. I'm nervous about the crossing for all of us. First, though, we must find the big herd, and so we follow the river south. After a long ride we inexplicably turn north onto the rough tundra again. "What's going on?" I ask. "I thought we were looking for the reindeer." Another sled comes next to ours: "The reindeer have already crossed. They're on our side of the river now," Nikolai says. Enormously relieved, we bump north toward the calving ground.

During a rest stop, our lead reindeer spooks at something and starts running. One of the others falls inside the hames, tangling his back legs in the lines, and is dragged. Finally, they stop and Arthun untangles the mess. We get back on our sled. Where the snow road ends, we detour into melting bogs and cross a small lake where the top layer of ice is covered with standing water. Reindeer splash through, their legs lifting high, water spraying on both sides. There's a jolt and the sled drops to a second icy layer.

Candle ice, hanging in slender fingers at the edges of ponds, pings as we pass. We slide up to a hump of dry land and back down onto ice-glazed snow, the sled runners clattering. Sun sparkle blinds me, and for a moment I'm lost. In a nomad's life there is no respite from movement, no anchored center toward which or from which to flee. A hard jolt brings me back: We are traveling downstream to find the reindeer, flying across frozen ponds and splashing through the ones that are melting; a second hard jolt dumps us upside down into pond water.

I call out for Andrew. His face is underwater, but he's holding the camera straight up in the air. Red Beard hasn't noticed that he's lost his passengers. We leap up and run after him. Finally, he stops and we get back on, dripping. There are snow clouds in the air. In polite Russian, Andrew asks Red Beard if he could slow down a bit, to which he growls: "You can hike if you don't like it."

We glide onto a good-size island amid melting ponds. Then we see: The big herd of reindeer are there, 2,500 animals swarming the hillocks and swimming the narrow channels. Something black moves between a few of the animals' legs. It's a newborn reindeer, tiny and wobbling. The mother is lying down in the midst of the moving herd: The calf is trying to suckle. Reindeer are running past. The mother gets up, lets the calf suck, then joins the others, with the calf running alongside.

One of the herders has gone ahead; we hear only his high-pitched, haunting, two-note call. He's trying to get the reindeer to cross a narrow bridge, but the animals shift nervously in a wide circle. An hour goes by, and another. It begins to rain. A crane flies up, along with three geese. At last the animals funnel over the bridge. On the other side is a surprising view, not tundra but a narrow strip of arable land, a hay meadow above the raging river.

We camp at the edge of its three Van Gogh–like stacks. There's a tiny hut with a long table inside but no stove. We pitch two tents—that's all we have.

Andrew and I are wet at day's end; rain turns to snow. Over a roaring bonfire the men whittle the ends of a forked stick and roast reindeer en brochette. As snow falls, we cram into the hut, drink shots of vodka, and eat berries and meat at the long table. There are ten of us, and even without a stove, our combined body heat keeps us warm. "The reindeer want to go north. They'd be at the Barents Sea in a few days if we let them go," Vasily says. "But we want them to stay here until all the calves are born. This is a good place for them."

During the night, wet snow pushes at the tent. In the morning the sky is black over white-topped haystacks. We drink pond-water tea. It is brackish, having been "sieved" through furled lichen leaves. Breakfast is reindeer soup. The bones are fed to the dogs. Red Beard's gruffness softens as he begins to sing. I don't know why, but I ask him if he's ever been in love. "Yes," he says quietly. "I once had a wife. I have a son. They've never been here to see me." Lines from a poem by Anna Akhmatova float into my head: "She whom people call spring / I call loneliness."

MORNING. We travel to the reindeer, half a mile away. Two more calves have been born. They are black dots in a swimming sea of animals. Soon there will be more than a thousand calves. In the meantime the men are roping young reindeer out of the herd to put them into harness for the first time. In the roiling herd the men, carrying soft ropes, run like gazelles. Katya's younger brother wrestles a young animal to the ground, slipping on a halter and dragging the terrified animal to a willow bush, where it is tied.

The herd is stirred up and running. Baby calves wobble, trying to keep up. They can run within a day of being born. When the animals stop for a moment, a calf tries to suckle, but the female strikes at him with her front foot (apparently it's not her calf) and the young animal is lost in the swirl.

At last, both the herders and the herded need rest. The men make chai-pi on an open fire. Whole slabs of reindeer meat are laid directly on the coals—a leg, some ribs, another leg. The meat sizzles and drips grease into the flaring fire. We cut off pieces from a leg, seared on the outside, raw on the inside, and pass it around. The men joke about how the smell of meat will attract bears and wolves. Bread is cut from a loaf and passed. The exhausted reindeer are bedded down in a wide circle around us. "We eat them but they still love us," one of the herders says.

I ask Vasily how he was selected to be chief. He answers shyly: "Alexander was chief before me, and when he got tired of it, the men got together and asked me." He says that they are here on the tundra, herding reindeer, only to help their aging mothers. When I ask him what he will do when his mother dies, he says, "Maybe I'll stay in a small chum with the other men and live outside like this and care for reindeer. It's all I know how to do."

THE TUNDRA IS MELTING. It has opened like a cadaver, all its secrets exposed. We've said goodbye, left the big herd, and followed the villagers' road past the neat haystacks. We turn east to find our newly erected chum and bump so fast over bulging hummocks of matted moss that my teeth chatter. We drop down into ponds where, a few days ago, there was ice. Now the water is so deep it comes up over the tops of the sleds, and when we emerge, the runners are threaded with duckweed.

This is a normal spring melt, but deeper down is the insidious melting of permafrost that stretches all the way from the Kola Peninsula to the eastern shore of Chukotka and the Bering Sea. Thermal expansion of water is drowning islands and peninsulas, and lakeshores are being inundated. Critical nesting sites for the Siberian crane are being lost. Tundra soils that have warmed up to above 32°F. are collapsing. Snow, once liquefied, flows into ground fissures—wedges that formed thousands of years ago— and melts out huge tunnels of subterranean ice. Then these collapse, leading to more fissures.

Warming temperatures are endangering reindeer health. Late autumn or early spring rain falls on snow, drilling through the white cover and pooling on the ground, then freezing and forming a surface too hard for the reindeer to penetrate with their hooves.

Pollution, especially airborne soot from the burning of coal and from diesel engines, settles on snow, causing it to melt faster because the soot is black and decreases the albedo. Spring run-off can come a month or two earlier than it used to. Herders who used ice to cross from summer to winter pasture find they are trapped by high rivers and cannot cross at all.

Methane fluxes—there are an estimated 15 million tons of methane emitted per year from the vast storehouse of the frozen ground—are the biggest threat to the tundra. "The tundra is killing itself," one Russian scientist said. "It is sending out more methane than it is taking in. It has to freeze for the ecosystem to work. The tundra is turning green."

Spring has turned to summer weather. The ground thaws. We head to our new camp, a small chum near Snopa village put up by the men who are not tending the reindeer. On the way we pick up a few stray animals and splash through slush that quickly turns into running streams. Ponds have come into being that were earlier covered with ice and snow, and watery moats have widened. Water flies across the seat of the sled. The reindeer run fast. They know they're going "home," and the livelier they are, the wetter and bumpier the passage. But the sun is out and we're laughing.

"All that exists lives," the Chukchi reindeer herders say. Their compass has 22 separate directions, and the sun is a man walking, wearing bright garments. His wife is called Wandering-around Woman. For the people who live with reindeer, diurnal and seasonal movements are continuous yet intricate—full of life—and we feel that too, traveling with the male reindeer that have been separated from the calving herd.

Near the snow road that leads to Snopa, we enter the men's chum. It's small and dark inside because there is no window, and the low table where they eat is strewn with dirty dishes, empty vodka bottles, and half-eaten bits of reindeer. The fire has gone out and the door to the stove has been left open. Everyone is gone. "Where's Fyodor? He was supposed to look after things," someone says. Arthun looks toward town with a mild look of anguish. I step away to stand in the middle of the "highway." It is five feet above the tundra and packed hard. Here and there a rivulet cuts through, forcing reindeer sleds to go around. I pace up and down. The road is a white string that leads to vodka on one end and in the other direction, who knows where?

Nikolai and Arthun walk to town. They don't want to be left out of the fun. "We only get near a village once a year, so we go in," Vasily says. A dog cries. It's Mu Mu. Red Beard has gone to town and neglected to feed her, so I give her scraps of meat. That night, for the first time in almost a month, I sleep alone in my own tent. In the middle of the night, using hot water from a thermos, I treat myself to a sponge bath, put my long underwear back on, and stretch out in my down bag, happy.

Men's voices wake me. Nikolai and Arthun return from Snopa drunk, but Fyodor is still missing. One of the older herders sends them back to fetch their friend. A few hours later, they return again empty-handed. "I'm afraid that he may be getting married tonight," Alexi says, "and we need him!" It's late when the rest of the men come back. There's loud talk, drunken arguments, and yelling. I pull my cap down low over my face and sleep.

In the morning squabbling ptarmigan wake me. The chum is in total disarray, and the men are asleep with their clothes on. Reindeer graze nearby, and despite the spring warmth, the snow road is still intact. Then I see Fyodor in the distance, walking bandy-legged toward the chum. He's alone. There's no woman with him, but he's carrying a puppy.

"Guess what?" he says to me. "I can speak five languages now. Do you want to hear them?" I ask him to speak English. He cocks his fur hat and comes closer, his face almost touching mine: "Mac, mac mac, mew, mew, mew . . ."

"What's that?" I ask. "English," he says, grinning. Stepping back, he spreads his arms wide, dropping the puppy carelessly. "I was almost married last night!" he says. "But Alexi broke up the ceremony." The dog disappears inside the chum. "Oh well. Next year when we are camped here by the side of the snow road to Snopa, I will go again to the village. She'll be there and I'll ask her to live with me on the tundra."


PINE WIND. Ten swans circling. Partridge noise in the brush. Of all the men, Arthun, the youngest, is the most responsible. When he goes out to check the reindeer, I ski up and down the snow road in search of bear tracks, wolf tracks, or incoming swans. "You have to stay here for a whole year before you know how we live and who we are," Arthun says as he passes by in his reindeer sled. "Yes, you are right," I say.

Later, a horse-drawn sleigh driven by a Nenets man, with his young daughter beside him, appears. They're bringing Red Beard home. "He was too drunk to walk, and he might have frozen to death out here," the man says gently. Tundra etiquette means that both villagers and reindeer herders look out for each other. We invite them in for hot chocolate and tea. The little girl looks terrified and I try to console her. In his drunken state Red Beard sings, then falls back against a bedroll and begins snoring.

In the morning we pack. Fyodor is missing again, so Andrei, Gordon, Andrew, and I are asked to help bring in the reindeer. The men are ashamed to have to ask for our help, but we're secretly delighted. We drive the animals toward the portable corral and push them forward until they hit the far end, then mingle, antlers clacking. One by one the men rope them out of the herd and take them away to be harnessed.

The sky is clearing, but it's cold. Ice spangles the net enclosure. Spring has come earlier this year, the herders say, and they worry that the mosquitoes will soon arrive. It's time to go. The helicopter is scheduled to pick us up at 11. We pile our gear on a waiting sled and ski behind it to Snopa to wait.

Loose village horses are running. Tails flying and heads high, they are prancing and snorting. One thick-legged gray workhorse bucks across the field in front of us. We're on what the villagers call the airfield, an empty, snow-covered pasture with a tiny shack at one end. The reindeer and the sled are parked under a limp wind sock. While waiting for the helicopter to arrive, we tour the village. It is a handful of wooden houses from the 1800s, with beautifully carved window frames and lace curtains. Each house has a greenhouse and a garden plot. Horse-drawn sleighs carry milk from the dairy. There's a communal wood-fired bath and sauna and fresh bread just coming out of the village bakery. In the schoolhouse, the teacher proudly points to the portrait of Pushkin that has replaced the one of Lenin. A small store sells cigarettes, hard candy, packaged cookies, and vodka. By morning most of the men in the village are drunk.

Nikolai and the others have been searching for Fyodor. He was seen running between two barns but quickly hid. He knows he's in trouble because he didn't come back to help gather the reindeer. Later, when he shows himself, the herders shun him, and he mopes off, no longer a "happy drunk."

We lie on the ground, our heads on our backpacks, looking skyward for the helicopter. Arthun gives me a gift of a reindeer's headstall made of leather and bone. "Everything we make is done with a knife and an ax," he says proudly. "Soon we will be away from this place and there won't be anyone getting drunk." We stroll to the cliff overlooking the Snopa River, the one we were to cross just a few days ago. Now the ice has melted and the water runs brown. Arthun says, "I can hear water running everywhere. The reindeer's antlers are growing back, and today I saw four geese, that's how I know it is spring."

Later he says, "I'm sorry you had to see us this way, with some of us getting drunk. It's usually not like this. We live far from villages. I wish you could stay. Yes, a whole year would do it. Then you would know us, and after, you could go your way."

The helicopter lands in the bare field, a beast falling out of a nightmare. There are passengers inside who have come from other remote villages and are headed for Arkhangel'sk, women wearing helmetlike hats and thick wool coats. As we board, the herders hand up our skis and duffels. We stack them on top of the other loose baggage in the middle of the aisle.

The door slams and the aircraft lifts with a shudder. Tears come to my eyes. I will not be here again. Cloud shadows sprout and die. I turn my head to look: It is endless, this mosaic of rotting ice and gray-green lichen. As we gain altitude, I'm instantly lost. There are no villages, no chums, no reindeer, no residence-on-Earth simply forged by bone, wood, hide, and ax. High up in the sky, I understand that the body of the tundra is bigger than all that; it dwarfs human and animal occupation. All I see are islands, hummocks, and meltwater moats, and a whole frozen world going soft in faint light. I strain to hear the lichens' and fungi's symbiotic song and dance, fearing that the beat will soon be a cranberry bogs' cough, methane boiling up in a million pond pots.

We hover, then slide over a world of leaking vessels—as if taking in a last breath. Is the Earth bleeding to death? Trees, when they appear, stain the sky with shadows that are blue, not black. Geese drop down onto water (or is it sky?) and swim against reflected cloud drift. Lake ice is gouged by thawing reindeer tracks. Collars of ice fold back from waterways. A tree goes down—I see it falling. Beyond, five swans land. A bend in the river dangles like a loose knee.

We land at Oma, then Mezen', then at a nameless village. As we pass into the world of trees, I look back at the vast expanse from which we have come and see that, once this generation of Komi are gone, there will be only the wiggle-worm puzzle of melting tundra, a whole planet brined.

# IN A DARK LAND

>>> <<<

A dream before flying north to Nunavut:
*Someone is dying. He's calling my name. We're in a dark place. A single eye is roaming the room. He asks me to grab the eye and put it in his forehead so he can see his own demise.*

DECEMBER 6. Iqaluit, Baffin Island. Falling farther and farther from sun.

In failing light I board a small plane with insufficient heat and sit behind a teenage boy who is on crutches. He wears only a thin windbreaker, one shoe, and no socks, leaving the toes of his broken foot exposed. The young bush pilot has been staring at him, "You ought to wear more clothes. If we went down in the tundra, you'd be first to freeze." The boy laughs, then, sheepishly, makes the sign of the cross as we take off.

We are headed for Igloolik, a town on an island of the same name at the north end of Melville Peninsula in the eastern Canadian Arctic. Nunavut's Arctic Archipelago, with its scattered islands and narrow straits, is more land than ice. Subsistence hunters here say they are "going out on the land" to hunt instead of "going out on the ice," even if their prey is a marine mammal like walrus or bowhead whale, harvested on the sea ice. To the northeast, where Greenland rises mountainous and topped by an ice sheet, travel and hunting is almost exclusively on ice.

The last pink of a short day fades as we fly northwest across the ice-scoured rock of Baffin Island. In December the top of the Earth is tilted away from the sun, and at latitude 69° N, Igloolik will have only a few hours of twilight—no direct sun. Over the years of Arctic travel, I've come to enjoy the trance of perennial night, but today my mood is somber.

A howling wind buffets the plane. We fly straight into it. As night comes on, we rise above clouds into a catapulting darkness that pushes into every corner of the plane, tightening and loosening the psyche like a vise. Ahead, under clearer skies a moon glint emanates from ice. Twenty thousand years ago all this was covered with an ice sheet. Now, with that burden lifted, isostatic rebound is occurring on these gravel and bare-rock islands: Igloolik is rising three feet per century. A continental climate keeps Nunavut frigid in winter; its island archipelago and narrow straits hold sea ice firm even as other parts of the Arctic are losing their sea ice.

The dark time in any Arctic village is quiet—not much hunting activity but plenty of meetings, planning, school functions, and equipment repair. A hunter on the plane, coming home from a meeting in Iqaluit, reminded me that in the early times, Igloolik was not a town at all but one of many seasonal camps where seminomadic families hunted abundant walrus, narwhal, and seal. Caribou were nearby—they crossed the small strait to Melville Island and went inland. Even now, hunters wear caribou anoraks and leggings because they are exceedingly warm and weatherproof.

"Mostly we traveled," the hunter says. "Across to Baffin, up to Pond [Inlet], west to Boothia, south to Melville. Animals took us where they wanted us, summer and winter. If we wanted to eat, we followed."

The plane shivers and drones. We fly across the thick waist of Baffin Island, then north along the edge of Foxe Basin. As I look out the plane window, the upended plates of rough ice near shore are sometimes visible, but very little else demarcates land from ice. From a volume of oral histories called *Saqiyak,* I read this: "We used the land as a tool back then. If snow was drifting, we knew the wind. We knew when blizzards were coming even when the sky was blue. Clouds told us wind was on the way. Rough water signaled land; calm water warned of drifting ice. The ice used to tell us things too. We watched it all the time. Long, wide cracks in the springtime meant the ice was thick and wouldn't melt soon; many small cracks meant it might come apart under you. If warm weather came too early, then summer would be cold; if heat came late, the rest of the summer would be dry. We were weather-watchers. Had to be."

The plane changes altitude, and we drop through a raft of clouds. Still, there are no village lights or signs of life anywhere. *Nuna* means "earth." Nunavut is the 770,000-square-mile territory claimed by and awarded to the indigenous people of Arctic Canada in April 1999 after two decades of negotiation. The founding notion of Nunavut was what its Inuit citizens call *qaujimajatuqangit,* or IQ, meaning "traditional Inuit knowledge."

The new Inuit government's mandate was to reflect directly its people's traditional values, knowledge, and worldviews by replacing the imposed colonial governance from Ottawa with one that would have an Inuit majority. Inuktitut, their language, was to have equal status with English. Control of lands, lives, and resources, including minerals, education, and wildlife, was to be returned to Inuit citizens. Core traditional values and cultural sustainability were key.

That was nine years ago, and the "old ways" and the words that articulate it are said to be quickly eroding. "There is no 'IQ' here," an Inuit friend says. "Despite everything that is going on in Igloolik, the accumulation of thousands of years of culture is fading away."

In 2006 I visited Sheila Watt-Cloutier at her quiet house on the outskirts of Iqaluit, the capital of Nunavut. A tireless spokeswoman from Nunavik—northern Quebec—she was nominated for a Nobel Prize along with Al Gore. She headed the Inuit Circumpolar Conference and now travels the world with her message about the importance of Arctic peoples in a time of changing climate.

We drink peppermint tea and look out at the icy bay.

"We think our culture is valuable and good for everyone to know about," Sheila begins. "We've lived at the top of the world for millennia without depleting a single animal or resource. It says something about the sustainability of our culture, and it speaks to the great disconnect that we are addressing up here. We are part of a very unusual place, and we are part of the world. We've changed from a powerful, wise culture to one that is self-destructing. We need to be as conscious as possible about controlling our own scene and to recognize what has wounded us and to keep track of who we are. It's not about keeping to strict tradition but to be able to choose how to live."

She shows me a photograph of her daughter, who is an expert throat singer and tours Canada giving concerts, an example of taking something from tradition and giving it a modern twist. Sheila holds the photo in her lap and sighs deeply. Flying the world tires her. "I love it here in my little house. It's my refuge," she says, then begins again explaining the lost past.

"When the cultural authority of the shaman was eliminated by missionaries, when our sovereignty was dismissed as unreal because we did not practice land ownership, a culture of dependency resulted. When you lose the power of the hunt, you lose the power of belief, language, and narrative that drove us. When we no longer live on the land properly, we lose our dignity. The wisdom of the land and the hunt builds character skills. It teaches patience, courage, sound judgment, and to be bold under pressure. These are qualities people need everywhere—these traits are transferable. The hunting culture is a modern tool kit to adapt to the world. There is no better measure of genius than to survive here, to be hospitable to what others think of as an inhospitable climate, to become one with the place, not to conquer. We Inuit embrace the cold. I don't ever remember being cold, just happy."

THERE'S COMMOTION on the plane. The boy on crutches and an older woman are excitedly pointing out the window at lights in the distance. "Where's that?" I ask. "Hall Beach," they say in unison. Then, straight ahead over the pilot's head, more lights appear in a frost haze. "Igloolik," they say. The name means "a place of many winter houses."

It looks too big to be an Inuit village. "How many people live here?" I ask. "Too many," is the woman's sullen reply. On approach the island glows bright. "Sixteen hundred," she says at last, just as the plane bounces down. "If you'd come last month, you would have seen the sun. Now you won't see nothing. The sun left us on November 29th and won't come back until January 14th."

The runway is perched on a flat bench above town. Below are rows of houses lining a horseshoe-shaped bay covered with ice. All is white. A few houses are festooned with Christmas lights. Skiffs lie on their sides, abandoned for the season at the edge of the frozen shore.

Snowmobiles are the source of liveliness: They arrive at the airport to pick up passengers. The boy with the crutches leaps on behind a driver and zooms away. No one comes for me. Soon I'm the only one left in the tiny waiting room. An older woman asks if I have someplace to stay. It's late, and they want to close up the airport and go home. I say yes. But I don't know when he'll arrive. Finally, a young woman appears and says, "John hurt his back. Can't drive. I'm your ride."

I've come to visit John MacDonald, founder of Igloolik's Oral History Project, in existence for more than 20 years. Born in Glasgow, Scotland, John grew up in Malawi, southeast Africa, but made his way improbably to the far north, where he met his wife, Carolyn. They lived in various Inuit communities but landed in Igloolik, with John heading the Igloolik Research Centre for visiting scientists. Carolyn founded an early-learning center and a Head Start program.

Fluent in Inuktitut, both John and Carolyn are knowledgeable about Igloolik's material and intellectual culture and stalwart members of the community. They hunt and fish, build kayaks, and sew traditional fish-skin pouches; they know the legends and the way-finding words. But the median age in Igloolik is 18, the unemployment rate is nearly 80 percent, and the town contributes its share of suicides to the Nunavut rate of 77 per 100,000 people. John and Carolyn have choices others here don't have. They can always leave and go home. "Except," John says, "this *is* home."

When I met John here a year earlier, he told me about his Oral History Project. It began in 1986 as a collaboration between the Inullariit Elders Society and the Igloolik Research Centre that John headed. The aim was to collect all that remained of the traditional knowledge of the Amitturmiut—the Inuit people of the northern Foxe Basin. There are 600 carefully translated oral histories of elders in the area, each one archived and key-worded on computer software at the lab.

"It's always too late to begin such a project if you start thinking of all that's been lost," John says. "On the other hand, it's never too late to begin." He found that even in the 1980s, there were still plenty of people alive who had lived traditionally and were eager to pass on what they remembered from those who taught them—parents and elders—as well as the traditional ecological knowledge they knew themselves, firsthand.

The town lights are a kind of sun, but one that has been eclipsed. I take off my winter boots and climb the stairs. "I hope you aren't coming up here to ask about climate change," John says, first thing. I gulp and say nothing. He's tall, lean, and white-haired. "Because we're not affected by it. I'm tired of the crisis narrative about global warming. It's a lie conjured up by the southern media [i.e., in Ottawa] that diverts resources away from our real problems. What we have here is something more immediate. We have social problems, a terrible malaise. The immediate threat is to the Inuit people here—their elegant culture, so brilliantly adapted to the world of ice that has for thousands of years kept the barbaric at bay."

His blue-gray eyes soften. He offers me a glass of homemade wine. Red, from French grapes. He takes a sip and smiles despite the fact that he's in pain. It's late and Caroline is already asleep. The sky is dark and the town roads are white sheets. Snowmobiles roar by.

I think of what Inuit hunters in Clyde River, on the east coast of Baffin Island, have said about the climate. They call it *uggianaqtuq*—"a friend acting strangely." But they are more exposed to open water and the variability that comes with a warming ocean and hard winds, whereas the narrow strait that divides Igloolik from the west coast of Baffin Island is narrow and the winter sea ice protected.

Too tired to argue, I find my bed in a windowless basement room. It's cozy and warm. John's outburst about climate change puzzles me. Eight months earlier, when I visited him here, he seemed less angry, but as the week progresses, I begin to understand the urgency of his appeal.

MORNING. Over porridge John and I talk. "The point of loss for the Inuit people here," John says, "was when they were moved into town. That's when things changed radically. Leadership was undermined, the skills needed to live on the land that had been learned and passed on for thousands of years, the complexity of the language and the knowledge grew weaker. The young people who have become true hunters can be named in one breath. It's becoming an unusual choice. Revered, but unusual."

Out on "the land" extended families lived seminomadically, moving from camp to camp, following the ice, the weather, and the animals. Home was anywhere there was food and family: Skins were prepared, clothes were made and mended, the walrus-blubber lamp—the *qulliq*—was kept lit, and children were raised. Each group had internal allegiances. How could they understand the instruments of authority that came from the outside, from the governing structures in the south? No one looked up to these leaders. They weren't the decision-makers from within family groups. "The change happened only 50 years ago, and the people here are still in culture shock," John says.

On the radio Christmas carols are sung slightly out of tune in Inuktitut. "The lure was, at first, schooling," he tells me. "They forced the children to come into town to get an education, and eventually the parents followed. There was also the lure of medicine and religion. A Catholic priest told them that if they confessed, they would be absolved of their sins. This pleased them. They could step out of their taboo-ridden society, toss away their amulets, and do whatever they wanted to do," John says.

Both Catholic priests and Anglican ministers forced people from the Igloolik area into baptism, pressuring them to say yes in what appeared to be a competition for parishioners. The priests went to the camps and told the people that God would give them everything they wanted, that God's power was stronger than the shamans'. As people were brought into a sedentary life—nomads have always been mistrusted by the greater society—the authority of a centralized, single God mirrored their relocation to town and replaced the lively pan-holy, spirit-glutted land.

Clocks and calendars were brought in by the missionaries so that people would know to come to church on Sunday. But hours and minutes do not make sense in a place where the year is divided into dark and light, ice and open water, where going out to harpoon a walrus might take days, weeks, or months.

The societal rules of a colonial government were at odds with the naturally egalitarian one of the Inuit. "Any claim to land ownership is anathema to the Inuit culture," John says. "Just the naming of land, giving it a single name like Baffin Island, did not make sense, since the land of Nunavut is known by thousands of seasonal place-names."

Yet Igloolik, only one of many seasonal outposts that fanned across the top of the Melville Peninsula and Baffin Island, became a permanent town. John insists that the government was never malicious in moving people to the community. I look at him skeptically. "It was, I believe, well-intentioned," he says. "When

Carolyn and I came here in 1985, there were still six or eight out-post camps occupied year-round. There was no coercion to move. It was simply not possible to take medicine and education out to every camp, and if a government makes no attempt to provide education and medical care for each of its citizens, it is culpable. How can you deny one part of a population what the rest of us have? Other countries have exterminated indigenous populations. This is, at worst, a misplaced paternalism."

John pours more coffee as the radio erupts in a fast-paced Inuktitut "Jingle Bells." Tapping my foot to the music, I imagine Igloolik when it was first established as being tightened like a purse, its contents centralized to the point of being crushed, and laid over the scattered fragments of the old. But the "old" keeps seeping up through the bone layers and middens preserved in ice and now loosened by melting permafrost.

"Jingle Bells" turns to "Silent Night" as snowmobiles roar by. "The elders insist that it's impossible to live a traditional life in town," John tells me. "They call present-day Igloolik *Qalunaa-nit*—the place where one goes to shop for white man's goods. No one ever intended to live together in one large group in a single hamlet."

Theo Ikummeq stops by. He's small and sharp eyed, wily and articulate, a hunter turned cultural spokesman who has traveled with the explorer Will Steger and teaches young locals how to hunt.

"We are a culture of walrus and moving ice," he says. "In other places there are no words for these things. The environment governs language and culture. Social structures here are not like those at Baker Lake, because each is governed by a different environment. Ice here, barren lands there. Lichens and caribou there, walrus, seals, and ice here. Now, for young people, our worlds are not many; they are all blended. The elders have a tough time passing it on. Our children don't need to speak Inuktitut to survive. That's scary. We are executing our culture."

Theo didn't move into town until 1978. He has nothing good to say about the place, yet he is as caught up in it as everyone else. In earlier times at camp, hunters used to play a kind of dice game made with the small bones of a seal's flippers. They'd throw them down on a flat stone. The bones that lay flat meant the hunter would get no prey; if any bones stood up, it meant they would bring home food. Now, in present-day Igloolik there are two gambling houses, teenage prostitutes, drugs, and drug dealers. "We went from the Stone Age to the Computer Age in 50 years," he says, laughing. "When the hunters come back from camp, they say this is a white person's town. It's an unhealthy place. There was always a spring migration of families getting the hell out of here. Now only 30 percent or 40 percent go out. Sometimes there are 40 tents out there. It's a good feeling when this town is out of sight. And you don't have to go that far."

Despite the wrenching social problems, Igloolik is the cultural capital of Nunavut. Inuktitut is spoken here. It is the home of Isuma, the Inuit film company that made the astonishing *Atanarjuat—The Fast Runner*. A separate women's filmmaking cooperative is also producing films. There's an Inuit circus troupe called Artcirq that recently performed in Le Festival au Desert in Mali, and a new aboriginal television station—Isuma TV—that streams Inuit-made documentaries online. It's a Renaissance town enveloped in social collapse, a community whose underpinnings have grown as dark as these winter days.

AFTER BREAKFAST John leads me down snowy lanes between speeding snowmobiles. To the southwest are the low gravel ridges of Melville Peninsula and the northeastern corner of the Barren Grounds, home to thousands of caribou. To the northeast are the mountains and glaciers of Baffin Island. Between is Fury and Hecla Strait, named for the two ships that William Edward Parry sailed to this coast in the 1820s.

Igloolik Point and the coast of nearby Baffin Island have been inhabited continuously for 4,000 years. What we call the Northwest Passage is actually the traditional and much used migration route of Inuit people who traveled on ice from the northeast coast of Siberia to the Canadian high Arctic, eventually crossing Smith Sound to Greenland.

The oldest human sites here were built around 2000 B.C. by the Tuniit, a paleo-Eskimo people identified with the Dorset culture. The ruins of their houses line the beach in front of the village. Each raised beach ridge represents a different era, with its signature houses, boats, decorations, lamps, and tools. Whalebones and a few human skulls can still be found there. Marine mammals were so abundant in the waters of Foxe Basin that there was no need to travel far. And there were plenty of caribou on the adjacent Melville Peninsula.

Around A.D. 1000, during the medieval warming period, Thule people pursued bowhead whales in open water with newly designed harpoons and hunted seals on ice. To the already ancient culture here, they introduced dogsleds.

Ice was the natural obstacle in the high Arctic, keeping out anyone who did not know how to live in the extreme cold. As a result, a single circumpolar strand of culture spans 6,000 miles, from the east coast of Siberia to the east coast of Greenland. The story told to a child in Siorapaluk, Greenland, is the same story told to youngsters in Point Hope, Alaska. The dialect may differ, but the language and material and intellectual culture are the same.

"There's nothing else like it in the world," John says. "Which makes me wonder why, of all entities, the Nunavut government doesn't see the importance of the Oral History Project, of the preservation and the passing on of tradition."

A WHITE HAZE covers the town. Houses and buildings are snow blasted. Rime ice etches windowpanes; mist coagulates into drifting sparkle. The sky is still dark. In the legends of the old days, before light had been brought to the world, it was said that inland dwellers lit their forefingers and used them as lanterns, and a nomadic hunter named Aqikhivik claimed that everything was alive. He said, "When a caribou had been eaten, the meat grew again on the bones. The houses were alive and could be moved with everything in them, and the people as well, from one place to another. They rose up with a rushing noise into the air and flew to the spot where the people wanted to go. In those days also, newly drifted snow would burn."

Today, in Igloolik nothing burns except electricity. Christmas tree lights flash. When the haze parts, John points out the star Arcturus—Sivulliik in Inuktitut. It is associated with one of the circumpolar orphan stories told everywhere in the far north in which an orphan and his grandmother escape a tyrant and are lifted into the sky and become stars.

"From stars we tell time and location as well as legends," John says. He learned celestial navigation when he was young and uses it still. His book *The Arctic Sky* is full of Inuit star stories. Between the Bering Strait and Greenland the constellation of Orion has 24 names—"the linked ones," "travelers," "runners," "hunters," "early risers," "those who follow," and "steps cut in ice"—the names suggestive of movement and travel.

Once there were two hunters who were trapped on moving ice. One followed the star Singuuriq and was never seen again; the other followed Kingullialuk and made it back to land-fast ice, "That's how important it was to know the stars," John says.

Today *qilak,* the sky, is a black dome. In earlier times it was thought of as a canopy of hard material that housed those escaping the hardships of life, as well as dead souls, and had an interior filled with feathers. Moon-man and Sun-woman lived there with the stars of dead spirits. The dome also had natural openings through which a shaman could make a journey.

The sun's first appearance was greeted by children with half smiles. They had to use their left side to smile. The other, unsmiling side was an acknowledgment of the cold winter weather still to come. Young people ran from household to household blowing out the flames of the blubber lamps. New moss wicks replaced the old. A flame was ignited with a flint and a stone inside a *tartuaq*—a box made of walrus-kidney membrane. From that one flame, others got the fire to relight their lamps.

Future weather could be predicted by what was explained as the contest between the Sun-woman and the Moon-man. If the sun rose before the first new moon of the year, the weather in spring and summer would be dry and warm; if it rose after the new moon, cold storms would continue coming.

The magical and practical lay side by side in Inuit lives. "They didn't think of them as being separate," John says. Snowmobiles roar by as we walk. Some drivers are young, others have whole families aboard—a husband, wife, and child on one seat. Do they know the stars? If they left town, could they find their way home? "Some do," John says. "Most don't."

JOHN'S ORAL HISTORY PROJECT is located in a three-story blue box that serves as the municipal building. Leah Otak runs the project. In her late 50s, she has long, black braids and a weathered face. She's soft-spoken but determined. Two years earlier we had met at her mother's house. They were talking about camp life: how in spring there was usually plenty of food because ringed seals, walruses, and belugas swam up the newly opened leads. Now her mother is dead and so is one of her two brothers, but Leah is carrying on the traditions and teaching the young.

"Everything is going too shallow," Leah says, looking out the window at the graves on the white hill. Just below is the high school, and in the other direction are tracks that lead far out to Ikpik Bay and Fury and Hecla Strait, where a few young hunters have gone to net seals.

"Our language is what's wrong to begin with," she says. "And that affects everything else. Our language has gone so English. The elders speak Inuktitut and the youngsters speak Inuktitut translated from English. If they don't know a word, they fill it in with English. Even the teachers in the schools don't speak proper Inuktitut. It is not a full language anymore. And it's only here in Igloolik that Inuktitut is really spoken at all. If our language is in trouble, our entire culture is in trouble. You can't have a culture without the words to describe and name what is."

Place-names in Arctic cultures point to the attributes of a place and can indicate the area where walruses, whales, seals, or sea birds might be found; they help the hunter remember how to get home. Inuktitut words have a sense of action and liveliness, naming and indicating movement—the direction a star might take, where it begins its seasonal rotation and where it ends.

John and Leah complain that television has helped push Inuktitut into a decline. Though the community voted down satellite TV twice in an effort to curtail cultural erosion, another vote in 1981 allowed it in. One wonders if the language protection bill on the table in Iqaluit, demanding that signs and services be written in Inuktitut, will do any good. "The underlying thing in all language survival is that people need to want to speak it," John says. "Here, it's a language of cultural practice, and when those traditional practices are gone, how relevant will Inuktitut words be?"

Leah offers us tea. She talks about her mother. "She knew everything," Leah says. "She left all her skins to me when she died, so I have to find time between raising children and working here to sew skin clothes.

"We've lost so much. The women are forgetting how to prepare skins, how to make a pattern for skin clothing. We have to re-learn before it's all lost—not in school, they have their Alberta curriculum, but somewhere else where we can get the kids interested again. They think it's only the past. To be acceptable again, we have to be Inuit in our hearts."

Leah was born in a winter camp at Iglurjuat, eight hours by Ski-Doo from Igloolik, in a sod house insulated with heather stuffed into the walls and under the beds. "Everything smelled so good," she remembers. "I was lucky, I wasn't sent away to residential school. I stayed with my mum, and she taught me everything she knew. My childhood was the tail end of traditional living, when the men ate in one room and the women in another. Food was shared. A haunch of caribou was cooked, a bite taken with a knife, and then it was passed around. During times of open water we traveled by umiaq that had a little sail if there was wind. It could be fast. We went to hunt caribou by umiaq. In the spring we hunted seals. And there was always walrus in Ikiq—Fury and Hecla Strait.

"We had three different camps and stayed in each until we had enough food and skins. There were no doctors. A man in a neighboring camp had a boil, and we stuck a lemming skin on it to suck out the pus. We got aspirin from the Catholic priest.

"Shortly after I started remembering these things, my mother's oldest brother converted to Christianity. But I kept thinking how wonderful it must have been to live without hearing any government rules or any ideas about God. That must have been nice!"

A young woman shows up in the office, dragging a small child behind her, yelling at William, a handsome man who has come north from Iqaluit to translate some of the oral histories into English. He listens, says nothing, and she leaves. Leah shoots a hard look at him.

"Our society in Igloolik no longer reflects our hearts," Leah says. "It's not the Inuit way to break into people's houses and steal, but now that's common. It's not the Inuit way to be boastful or envious, angry, violent, or disrespectful. Those of us who have been hurt in the past now give our kids anything they want just to compensate for what they did not have.

"Discipline is very important in our culture. Without it, you don't survive. When we lived out on the land, we were getting seals and caribou all the time. The women prepared skins and made beautiful clothes. Boys learned by doing, not talking. The men provided lots of meat for us. We didn't tell people to 'have a good day,' because we knew about Sila—the power of nature. We knew the day was beyond our control. It was a joyous way of living. There was no time to fight."

William goes back into his office and puts earphones on. The wind howls and the world outside goes white. Leah says, "Lately I sense that we're at our worst, here in Igloolik. I'm doubtful that we'll ever get back to our strong days. We are not the strong people we used to be. When we were kids, we helped. Life wasn't only playing. They're not making children good helping people anymore. Boy, we've really changed." She looks out the window. "Now who among us is *silatujuk*—wise?"

John turns on one of the lab computers and shows me how to access the oral history archive. I punch in keywords: "Shaman," "seal," "dog," "foretelling the future," and up come the stories that have those words in them. "We're the best studied community in Nunavut," William says. "But what good does it do?"

Alone in my cubicle I type in the words "Animal life." A story comes up: "One thing for certain there are not as many animals as there used to be, especially seals. In the spring there are fewer birds—we used to have so many things like eider ducks, old-squaw, and arctic terns, including *Saurraq*. In the past there used to be more, but now there are less, in fact I miss them." (Louis Alianakuluk Utak oral history)

"Dogsleds": "In those days they used walrus hide for the shoeing and *Isajuk,* caribou antlers when they were frozen. Bone shoeing was used too from a bowhead whale. These were the most common shoeing. We used sod too, but bone shoeing was best on moving ice. We made traces and harnesses in the summer for winter use. But we would not *arrsuq* (scrape the fatty tissue from the thong) when the temperatures were at their warmest, only time they would do that was when it was raining or in the evening. The *pittuq*—the draught strap—had to be made stronger. If it snapped the man would lose his dogs especially if he was a slow runner as he would not be able to catch up to the loose dogs. If he was a fast runner, he would have no difficulty." (Louis Alianakuluk Utak oral history)

"Seals": "Then there are the albino bearded seals. It is said that if you saw one you should not hunt them down. They are known as *silaat,* their hair is all white, at its back is a diamond shaped dark skin part . . . the best you should do is look at it even if you saw how you would love to catch it, but you should not try. . . . It is also said that your life might be cut short. This is what was said by the people before us." (Margaret Sunaq Kipsigaq oral history)

"Bears": "There are huge bears. From what I heard they are *inukpa-sukjuit,* giants. They are called *nanurluit.* It is said that if you are close to one of them and it blows out, this alone has the capacity to blow away a person . . . if it starts to inhale the person would get immediately sucked in. At least this is according to legend. Indeed, they are said to be huge." (Margaret Sunaq Kipsigaq oral history)

"Foretelling the future": "The wife of my Uumatiqati, she has the capacity to foretell weather conditions. And when they are going to lose a relative, they usually find they are having very difficult time catching a game animal, very much so. . . . Or it might be that a man wakes up in the morning, as he gets dressed, in the process he puts on a garment the wrong side. This he uses to expect something good to come this day, a catch of a game animal." (Lucien Ukaliannuk oral history)

"Animal transformations": "Sometimes animals take the shape of a person. I'm not certain how it happens, something extraordinary, something to tell others about something—this is *inuruuqajuq.* It may be a dog or another species of animal. You would first notice it as an animal, next thing you notice, after you turned away for a short while, then you see a person in its place. Sometimes the animals get the capacity to talk, at least that's what I've heard, they could talk. They were not feared. They were not scary at all. But when a loon uses its dance-sound in mid-flight there is something amiss. There are birds that fly in circles above you and they are chirping, this too was not liked by people. They know that something unpleasant is going to be heard. There were things that made you to know things, bad things, to happen." (George Agiaq Kappianaq oral history)

"Gratitude": "At the time when I was still too small to go on hunting trips and we were living in the igloo, the windows were frosted, including a frost built up on the panes at the entrance. When the frost starts to melt on its own, though the heat has not changed, my mother would immediately start to express her gratitude and announce that the hunters had succeeded in catching a game animal, and our supplies had been replenished. So it would come to be. She was able to tell from the defrosting of the windows." (Lucien Ukaliannuk oral history)

"Shelter": "This time of year we would start to move into *qarmaq* (sod house), while some would still be in a tent. Even when there was snow around, some would have stayed in a tent. Some would move into an igloo, using *tullaaq* (stomped hardened snow). Some would use slabs of ice hewn from freezing ice, *tugaliaq*, and some would have made a *qarmaq* before it got too cold. When you were in a tent in autumn, in the morning you wake up you will find your footwear frozen. Water in the pails would be frozen, then when the *qulliq* is lit, will slowly get warmer, so during the day the pails finally melt, that was the way it was. Everything was really hard in those days. It is now so different. Now we are controlled by money, in those days money was something we never thought of. The only thing in our minds was game animals and the need to survive." (Louis Alianakuluk Utak oral history)

WILLIAM EDWARD PARRY was given the task of finding a passage to Asia by the British Admiralty. The Napoleonic wars in Europe had ended, and Trafalgar, having grown bored, asked the navy to map the world and seek a shorter passage to Asia. Northwest Passage mania emerged quickly.

In 1824, on a second expedition to find the Northwest Passage, Parry and George Lyon sailed two ships, the *Fury* and the *Hecla,* up the wide waters of Foxe Basin. As winter came on, they anchored near the walrus-hunting village of Igloolik. By late summer the ice in the strait named for their ships was already three feet thick and the land was fog shrouded. For the next ten months Parry, Lyon, and their crews lived adjacent to 200 Inuit hunters and their families, going hunting with the men and taking Inuit "wives" for the season.

The Iglulingmiut, the local villagers, thought Parry had come looking for the remains of his mother, since their legends told of white people being the progeny of a marriage between a woman and a dog who lived on Qiqertarjuk, an island near Igloolik. The villagers helped Parry build an ice wall around the *Fury* to protect it, and drew accurate maps of the Melville Coast, assuring him that a passage existed, though, because of the ice, he was never able to find a way through to Asia.

George Lyon took to village life. In the *Private Journals of G. F. Lyon* he gave meticulous descriptions of the seminomadic hunters he came to know in scenes of both filth and splendor: "On the 25th of September 1822 I landed to visit my old acquaintances and found their huts in a most filthy state, owing to the mildness of the weather, and to their internal warmth: the water was dropping from the roofs, the ice had melted on the floors, and the juices of thawing and half-putrid walrus flesh, with other watery inconveniences, had made large sloppy puddles in the low entrances, through which we were obliged to crawl on our hands and knees."

Winter houses were made of bone and sod; *igluit* (the plural of igloo) were connected by low tunnels that fanned out like stars. They were built with walrus-ivory snow knives on the sea ice near the breathing holes of seals. Farther down the coast Lyon described a house made of freshwater slabs of ice: "Toolemak's dwelling was a perfect octagon and so transparent that even at some paces distant it was possible to distinguish those who stood within it one from the other; yet at the same time, it was so airtight, as to be completely warm."

Parry and Lyon enjoyed their new Igloolik friends and admired the ingenuity of their clothing: the deerskin mittens, double sealskin boots with walrus-hide soles, and "summer frocks" made of duck skins with the feathers worn next to the body.

Lyon reported that the women of Igloolik softened the bird skins by chewing them and stretching them on racks to dry. They made whalebone pots, ivory ornaments, gear for bows, fishing lines, and harnesses for dogs. "They also have an ingenious method of making lamps and cooking pots of flat slabs of stone, which they cement together by a composition of seal's blood applied warm, the vessel being held at the same time over the flame of a lamp, which dries the plaster to the hardness of a stone," Lyon wrote.

Their kayaks were 19 feet long, with 64 ribs made of dwarf willow, small bones, and whalebone. Dogsleds were six feet long, made entirely of bone with walrus-ivory runners. When hunting walrus in summer, they hoisted their kayaks on a piece of drift ice near a herd of resting walruses and paddled the ice toward the sleeping animals. Harpoon lines were fastened to the ice, so when a walrus was struck, it could not escape. When the animal tired out, the hunter put his kayak in the water and lay low, and when close enough, speared the animal to death.

Social life was easygoing: Marriages, divorces, wife exchanges, and mutual infidelities were carried on in mostly amicable ways, in what Lyon called "extraordinary civilities." Although the paternity of children was never certain, men treated their mates' progeny with equal care.

Lyon also observed Toolemak, who was a shaman, as he went into a trance: "A very hollow, yet powerful voice, certainly much different from the tones of Toolemak, now chanted for some time, and a strange jumble of hisses, groans, shouts, and gabblings like a turkey, succeeded in rapid order." Later, as the trance subsided, Lyon remarked: "The voice gradually sank from our hearing at first, and a very indistinct hissing succeeded: in its advance, it sounded like the tone produced by the wind on the base chord of an Eolian harp; this soon changed to a rapid hiss like that of a rocket, and Toolemak with a yell announced his return."

According to Lyon, healing consisted of blowing on the diseased organs and open wounds. Newborn infants were swathed in the dried intestines of "some animal," Lyon reported, then washed in its mother's urine. Charms made of the foot bones of wolverines, the front teeth of musk oxen, the eyeteeth of foxes, and the bones and teeth of fish were worn. A string of miniature knives made of walrus ivory was worn to charm the weather. Parry and Lyon ate the local specialty, *igunaq* (fermented walrus), and, in turn, taught European dances to the locals. They bought dogsleds and dogs and made frequent forays onto the ice. But when Parry had an argument with one of the shamans and tried to kill him with an ax, it's said that the blade failed to penetrate the man's body. "We didn't need guns or axes. Our weapons were carried inside," Sheila Watt-Cloutier said about the story. The shaman was reportedly so angry he made the ice become so thick that no outsider was able to sail to Igloolik or penetrate Fury and Hecla Strait for 40 years.

AS SOON AS EXPLORERS and whalers from the British Isles and other parts of Europe and from America began plying the coast of Melville Peninsula and the low-lying island of Igloolik, flour, sugar, knives, cooking pots, steel sewing needles, and wood, plus rifles, ammunition, disease, and rum entered Inuit life. Some hunters took seasonal employment on the ships and used cash or trade goods to buy food for their families, to make up for the months they weren't out on the land. The accordions and square dances brought by Scottish crews are still practiced today, and square dances are called in Inuktitut.

With the good came the bad. Whalers took on seasonal "wives" who bore half-breed children soon abandoned by their fathers. Epidemics of southern and European diseases followed. Bowhead whales and walruses were hunted almost to the point of extinction by American and European whalers during the 19th century.

As soon as the whalers stopped coming, the marine mammal populations began recovering, but the march of outsiders continued. In 1910 a Catholic priest named Etienne Barzin established a chapel at Avaaja, a camp 15 miles north of Igloolik. Anglican Bibles were printed in the newly created Inuit syllabary. Before then Inuit had no written language. It was easy to learn and to read, and it helped spread Christianity.

With a wry smile, John says, "Theirs was a taboo-glutted society. When they found out they could come to town once a month and make a confession and didn't need to follow the strict rules of taboos anymore, they were relieved. It freed up their time. It was a shortcut to salvation."

Traders replaced whalers. In 1921 the Hudson's Bay Company opened a post in Pond Inlet, and by 1930 another post had been established in Igloolik. Fox furs were the rage in Europe. Subsistence hunters were encouraged to stop winter hunts and become trappers. The Hudson's Bay Company bought the furs in exchange for credit at the store, where they could buy food and "southern goods."

While these amenities certainly eased a difficult life, it corrupted the seasonal round of subsistence living. Families congregated closer and closer to the Hudson's Bay store, and soon enough, the game nearby was hunted out. Dependency on the fur trade increased, but when the fox-fur craze came to a sudden end in the 1940s, the hunters faced starvation again. They had lost track of the cycle of game as well as the skills and "second sight" of the great hunters who had come before them.

But all was not lost. According to the Danish explorer and ethnographer Knud Rasmussen, who came to Igloolik in 1923, shamans and elders flocked to him, telling him stories. "The 'old days' were not that long ago," John MacDonald reminds me. In village after village, camp after camp, Rasmussen talked to anyone who would talk to him, and most did, and took meticulous notes—3,000 pages of them.

They told Rasmussen that in all living things were forces that rendered them sensitive to the rules of life, and that these forces were found in names and in the soul. The soul gave the particular appearance to each being. In humans, the soul was a tiny human being, in the caribou, it was a tiny caribou, and so on. The *inusia,* the soul, was situated in a bubble of air in the groin. Any offense against the soul became an evil spirit. Evil spirits could be used by shamans to harm people who had disobeyed the rules.

Shamans were male and female, and sometimes couples who jumped easily between the spirit, animal, and human realms. They could cause death or they could cure, they could find out where the animals were, and they could find people who had gone missing on the land "by turning into spirits and go looking for them," one elder said. She begged Rasmussen to take her with him as he traveled from camp to camp by dogsled. She jumped on his sled and would not get off, and ended up living for a year with Rasmussen and his lover, Anarulunguaq, and her cousin, Miteq.

"We used to believe that animals all have spirits. Birds, lemmings, seals, caribou, all kinds of animals have spirits. Even weasels and foxes have spirit and shaman could move into animals' bodies," one of Rasmussen's informants said. Among the people, songs, amulets, and strict taboos were the spiritual and social tools to maintain a thriving society; modesty in front of the weather, animals, and one's relatives was mandatory.

The human alone was weak and powerless. There were ruling powers that could take or give life. Sila was the spirit of weather and intelligence, of consciousness and nature. Nuliajuk, or Arnaluk Takanaluk, was "the woman down there," who controlled the marine mammals and the seas. She was pictured in a drawing by one of Rasmussen's informants as a tiny, pear-shaped being with long, spiked hair sticking straight up.

The Iglulingmiut and the Avilingmiut believed that some of them were marked by Sila and held its power within them. Those born on days of good weather were called either *silatiariktut*, good-weather souls, or *silaluktut*, bad-weather souls. Those who carried the spirit could influence weather by going outside naked, walking around, and crying, *"Silaga nauk, ungass, ungaa?"* ("Where is my weather, where is my weather?") It would then begin to snow.

Wind had a spirit, and when it blew too hard, the shaman wrapped himself tightly in his clothes so the wind couldn't blow anymore. The spirit of the snowdrift was called Oqalorak. It ruled the sharp edges. The harder a blizzard blew, the more the spirit was delighted. Oqalorak sent storms down onto hunters and laughed at them.

The moon spirit was Tarqip. He lived with his sister, Seqineq, the sun. As a result of an incestuous relationship, in which Tarqip accidentally slept with Seqineq, the brother and sister lived in a double house in the land of the dead.

Helping spirits could be almost anything. One called Nartoq had a nose that protruded from its forehead, a lower jaw that was part of its chest. Its threatening behavior was meant to remind its "owner" that he was too easily angered. Another helper named Igtuk, meaning "boomer," made the sound of thunder in the mountains. "No one knows where he stays," Rasmussen's informant said. "He is made otherwise than all other living things: his legs and arms are on the back of his body, his great eye is just level with his arms, whilst his nose is hidden in his mouth. On the chin is a tuft of thick hair and below it, on a line with his eye, are his ears. The mouth opens and discloses a dark abyss, and when the jaws move one can hear booming out in the country."

Shamans talked to these and other spirits using special words called *irinaliutit*. The words could be bought and sold, handed down, or communicated as a legacy before death. The shamans, using these words, could cause dead things to come back to life. They could make frost appear. They could create life out of dead things and make clothes come to life. Nature and culture were bound together tightly, no seams between. Everything was alive, walking, hunting, singing, talking, listening, behaving, and "together-living," Rasmussen's informants told him.

George Aggiaq Kappianaq, born in 1917 near Igloolik, remembers meeting Rasmussen. He referred to him as a "half-breed white man" and recalled that Rasmussen's living quarters were very bright. "On the table there were cookies in a tin. I believe I might have helped myself as much as I could," he said. He was six years old at the time.

George was a frequent contributor to the Oral History Project. He wanted to pass on all that he knew and remembered of life out on the land. There were suicides even then, he said, and remembers finding a man who had shot himself because "he had too much in his head." He brought the man's head to his mother, who was one of many local shamans, so she could "wash it out."

George's life was filled with spirit helpers: "I remember seeing the helping spirit as a bird. It was this high [he raises his arm], and there was an aura. I could see it walking around, back and forth across the porch. To become a shaman you fast for five days—no food, no water. Then the person will make the sound of a bird.

"My mother became a shaman. Her helping spirit was that of a white man from the ship. She said she could see his aura as a bright thing from that person. Even in the dark she said that it became bright as day. A woman can be even stronger than a male shaman. Very much so. Some women were able to be a lot more powerful than men in shamanism. That's because of their mental state. They are more reluctant to harm others. For this reason they are more powerful."

George had a spiritual bent as well. He said: "I could see a beast right through the wall [of the igloo]—maybe a polar bear or a dog. I could even see the stars." He remembers doing what he called "going out with something other than a woman"—having sex with a dog, a caribou, or a seal. "I personally started to do it. If I had not confessed, I would be long dead," he said. "You have to confess to a shaman, because if you hide anything, you will get cancer and die."

About converting to Christianity, he said, "It's hard to tell if this religion is truly a good religion, because it leads to hate and despising. One must have love and practice it. One must not get too occupied with the forces you are expected to despise."

He was sure that the old ways still percolated through Inuit society, no matter how much they changed. He said: "There are still shamans around and always will be."

In 1923, Rasmussen said to those who doubted: "If in these myths are things which seem to be contrary to common sense, it is merely because the later generation is unable to grasp everything that, to their forebears, were obvious truths."

WITH PRIESTS AND TRADERS came schoolteachers and more missionaries, as well as "southern" medicines. After World War II the Canadian government decided to develop the far north less out of concern for the well-being of its Inuit citizens than concern that assimilation take place and Canadian sovereignty be firmly established.

When it was discovered that TB was ravaging the Inuit populace, a ship was sent to take people to sanatoriums for treatment. But it seemed more like kidnapping, a continual effort to subjugate these wild people. "They came for us at night," a former TB patient said. "We didn't know where we were going or if we would see our families ever again."

Sheila Watt-Cloutier recalls: "During the TB scare, people were moved out of their villages on a hospital ship. As a convenience the government gave every Canadian Inuk an identifying number, engraved on dog tags. Numbers were used instead of our real names. I'm EA3582. They were trying to erase us, and we didn't like it. Finally they put what they called Operation Surname into effect, giving us all Christian names and dropping the Inuit names by which we were known to each other."

In the 1950s residential schools were added to the litany of demands on these once self-sufficient subsistence hunters and their families. The eldest boy or girl of each family was sent away to one of two boarding schools—one in Chesterfield Inlet and one in Churchill, both run by priests.

Sheila recalled life in Nunavik, northern Quebec, before she was sent off: "I remember in my life a sense of groundedness and peace. Of control, of trying to capture the spirit of the old ways. The historical trauma that changed the course of the hunter and the wounding of the Inuit hunter has changed him to one who is struggling to find his place in a world of institutions."

"I was ten when I was sent away to school. They came for us in the middle of the night and we were put on chartered airplanes. Classes at these schools were taught in English. Inuktitut was not allowed in or out of the classroom. The curriculum was southern, with no reference at all to Inuit culture. We all had a number and a bottle of lice medicine. We were being reprogrammed. Many of the children, especially the boys, were subjected to sexual abuse. I didn't come home for five years. I grew up on a dogsled. I came home to a Ski-Doo."

Theo Ikkumek said that as a child he and his family were nomadic. "The first six years of your life are the most important. They determine how a child will be in the future. My education started in my first moment in an igloo. In the summers and falls we lived in skin tents and sod houses. These were my memories. Then I was sent to residential school at Chesterfield Inlet when I was seven. A few years later my brother Emil had a vision that kept me from going further with education. He kept me back to teach me what he knew. His vision held me here.

"At that school we were abused sexually. But getting some education enabled me to step forward. Some never came into the stream of things. Quite a few Inuit leaders have come out of these schools. Chesterfield was Catholic. I burned the pictures I had of those days. They knocked the school down."

Leah Otak says "healing has to be deep. In Igloolik we have lots of hurt people. I was married to one of them. The ones sent away to school at an early age to Chesterfield were abused, not just sexually, but also culturally. Now they are getting compensation checks. Big amounts. And they don't know what to do with the money. Family members hang around asking for things. We were never like that. There is so much sickness. *Qallunaat* sickness—white man's sickness. The healing should take place here, not at alcohol and drug treatment centers in Ottawa, because when they go there, they never come back."

Theo remembers the summer he returned from residential school and his older brother, Emil, took him out onto the land to teach him the old ways. He had to prove he had learned his way around. "I was 12 when I went out on the land by myself with the dog team and built an igloo, spent many nights, and went hunting for my food. The notion of boredom didn't exist. Even now, it's hard to comprehend. But the life can change too fast. We're lost now. Who am I? A Canadian? Just that question says it all. And most answer, 'Yes.' "

In the 1960s permanent towns were established: Arctic Bay, Pond Inlet, and Igloolik, where education, medical care, religion, and welfare were force-fed to all who came and stayed. Day schools were built and attendance was made mandatory. Outpost families were forced to send their children to town for education.

The children lived in boarding houses, but the families, so unused to being apart, couldn't stand the separation. Small, uninsulated houses with oil heaters, water tanks, and electric lights were provided. To pay rent, they were forced to accept welfare in the form of "Family Allowance," and still do today.

Seduction, need, fear of the consequences of refusal, and assimilative actions that were really enforcement—these small nations of subsistence hunters that had thrived for 4,000 years at the top of Foxe Basin and Baffin Island were suddenly confined, sedentary, overcrowded, overregulated, and hooked into a cash economy from which there is still no release.

The final blow came when the the Royal Canadian Mounted Police (RCMP), first stationed in Pond Inlet but now with a post in Igloolik, began slaughtering at least some, if not most, of the sled dogs. Without dogs, the hunters' imprisonment was complete; without dogsleds there could be no hunting, no outlying camps, no seasonal movements that followed the migration of walruses, seals, birds, fish, and whales.

DOGS WERE ONCE a part of everyday life across the Arctic. Though they did not always pull sleds, they were brought from Siberia to Alaska 10,000 years ago. Lucien Ukaliannuk recalled for the Oral History Project: "Litters born in the winter were kept inside so they would not die from the cold. We are alive today because there were dogs to help our ancestors to survive. They depended solely on their dogs to secure food and meet their needs. Indeed, I remember those days when we depended on dogs."

Women were often asked to raise and feed the young dogs. "They have the know-how, just like they have the experience of rearing children. She feeds them just right—small pieces at a time, not feeding constantly.

"Dogs are very knowledgeable. They might be hungry because there was no way to get game animals, like if there was no floe edge. So the dogs were *niriujaaq*—expecting something. Also, they can take you home. In those days, my dogs were a lot more observant than I was. There were times when I lost my direction, but my dogs knew where to go that would take me home. This was particularly true during a blizzard, a heavy snowfall, or poor visibility. When you stop your team the dogs will get down and settle. One among them will get up and *niugarsaq*—rub his back on the snow. The reason he is niugarsaq is that he's anxious to get home, he knows the way, so you put the longest trace line on him so he can lead the way."

During a walrus hunt on moving ice, which is very dangerous, a hunter would sometimes take an old, retired dog out with him. Lucien said: "He puts that dog on the shortest trace who hardly had to pull, because he will know which way the ice is moving and when to get off it before it breaks away and which way to go, because he can feel the currents and knows the wind."

When the government began taking over, they demanded that the dogs be tied up. Lucien said: "That is when they started to get really bad healthwise, they had to be fed while they were in chain. This caused them to get unhealthy very quickly."

If the dogs got loose, they were shot. "This person was going to tie his dogs as he was just returning, and before he got a chance to tie them, all of them were shot. When you are just returning from a trip, you would unharness them and the dogs run around before they settle. They will return, close to the dwelling of the owner. That was how the dogs behaved. When the government came to town, they started to shoot off our dogs," Lucien Ukaliannuk said in his oral history.

The Qikiqtani Inuit Association's Truth and Reconciliation Committee has been holding hearings on dog-killing issues. Allegedly, 20,000 sled dogs were shot in the 1950s and 1960s in Nunavut and Nunavik, and the Nunatsiavut region of Labrador. Government administrators say that is impossible since there was not enough ammunition allocated in those years to kill that many dogs, and that the few killings were for public health reasons. Inuit hunters say otherwise, and the controversy continues.

John MacDonald insists that it could not have been an all-out slaughter. "There weren't enough policemen to shoot all those dogs," he said. "I think it happened irregularly." But Sheila Watt-Cloutier disagrees: "There was a systematic slaughter of our sled dogs," she said. "Thousands and thousands of them were shot by the RCMP. It was an assault on us and our way of living. They even shot the dogs of some visitors to our village from another area who had come to trade, so those people couldn't get home."

Regardless of who is right, writers for *Nunatsiaq News* say that what's important now is to heal "the collective mourning for and unresolved grief at the loss of an old way of life."

Shortly after sled dogs became scarce, snowmobiles were introduced. They were expensive and used petrol. To have one, to be a hunter, to have any mobility at all, meant you had to become "a wage slave," an indentured servant to the government jobs on offer. The emotional and economic strings that had lured the hunters into town kept pulling tighter and tighter.

Family groups who had never lived near each other were crowded together irrespective of kinship lines. Ceremonial life disintegrated. The local shamans went underground or disappeared. Language weakened. A whole way of being and an ecology of thought vanished.

The only imperative these hunters knew was the one laid down by Sila. Weather and the movement of animals ruled. They had always been told to observe Sila and to be alert to the dangers of weather and ice. But where, in the exhaust of snowmobiles, does Sila reside?

IN 1953 TEN FAMILIES from northern Quebec and Pond Inlet were "relocated" to Cornwallis Island and Grise Fjord and the east coast of Ellesmere Island. This "solution" to what Canadian ministers were calling the Eskimo problem—the failure of the fox-fur market in Europe on which Inuit subsistence hunters turned trappers had become dependent—was actually a veiled attempt to claim sovereignty to areas of the far north and to prevent Greenland hunters from crossing over to Ellesmere, only 30 miles from them, and shooting what they claimed to be "Canadian" polar bears.

The relocation of these families a thousand miles farther north was brutal. Cornwallis is low, windblown, and gravelly. Parry had overwintered off the coast on his first Arctic trip in 1819, on the first of his three Arctic expeditions, and he had found nothing there but rock—no open leads where seals could be caught until spring. The east coast of Ellesmere Island had been uninhabited since the Medieval Warm Period, when Thule people hunted up the coast to the Bache Peninsula. Winds were well over 100 miles an hour and pressure ice heaved up in the narrow strait, making access to marine mammals extremely difficult and dangerous.

The relocated hunters were unused to such weather extremes and were not equipped to hunt on moving ice. Musk oxen and polar bears wandered by, but they were not a dependable food source. The Canadian government may have gotten rid of their "welfare Eskimos," but they sentenced them to a life of homesickness and near starvation. Their promises of supplies and a return home after a year if desired both proved false.

Among those "exiles" was the son of American filmmaker Robert Flaherty and his Inuit lover Maggie. Flaherty's romantic documentary *Nanook of the North* (1922) became a classic, but after making it, he abandoned Maggie and returned to his American life and family. His illegitimate Inuit son, Josephie, lived in spiritual poverty, the poverty of being dispossessed.

Young Josephie was kind and well-meaning, but he was neither hunter nor villager. His adoptive father, a fine hunter, was one of the first to be relocated to Ellesmere Island, and on his way there wrote a letter to his beloved stepson, asking him to join him.

Josephie had already packed up his wife and children and was on the annual ship north when he learned that his stepfather was dead. Once on Ellesmere, Josephie was unable to cope with the ruthlessly cold weather. While his biological father, Robert Flaherty, was being wined and dined around the world, Josephie and his family—homesick, grief-stricken, and despairing—began to starve.


THE RECENT NATURAL RESOURCE MAP of Nunavut that I am holding is laced with little red boxes and color-coded dots marking oil and gas, diamonds, uranium, coal, gold, rubies, iron, nickel-copper, and precious metals. A coal project on Ellesmere Island, an iron ore mine on Baffin, gemstones to the south of Igloolik, gold near Churchill, gold and uranium near Coronation Gulf. Between are musk ox refuges, caribou calving grounds, narwhal and walrus migration routes, nesting bird cliffs, and polar bear sanctuaries.

Issues of sovereignty are not over in the Arctic. Northwest and Northeast Passage mania is raging again as areas across the top of Russia and through the Canadian archipelago remain ice free longer. In the summer of 2007 the Russians planted a flag at the North Pole, drew a line from there to the tip of Siberia, and declared that roughly half the Arctic, an area that encompasses 6 percent of the Earth's surface, belonged to them.

Although the 1800s rush was only to find a shortcut to Asia, now every world power is trying to lay claim to enormous untapped circumpolar oil and gas reserves, as well as iron and precious metals. The entire Arctic is open for business, with no thought given to the consequences on the fragile Arctic ecosystem: its marine and terrestrial animals, humans, migrating birds, insects, plants, fish, and the open-ocean ecosystems. The climate crisis we are facing now results from human domination over the planet. It turns out it wasn't really "ours" after all.

There are eight proposed uranium projects in western Nunavut. Inuit protesters in Baker Lake say they would be downstream and downwind of the mine, and the nearby caribou calving ground would be adversely affected. The hunt for uranium has gone farther west to the Thelon Wildlife Sanctuary on the border between Nunavut and the Northwest Territories, where two separate caribou herds have a combined population of 772,000 animals.

There's a diamond rush on in Nunavut. More than seven million acres of land have been acquired by various aggressive mining companies on Victoria, Devon, Banks, Prince of Wales, Baffin, and Ellesmere Islands. On the western waters of the Northwest Passage, in the Mackenzie River Valley and Delta, there are an estimated 1.5 billion barrels of oil reserves and 9 trillion cubic feet of natural gas. A 600-mile-long pipeline is being built from the town of Inuvik south through Fort Simpson into Alberta.

Nunavut has a new colonizer: the mining industry. It believes the pursuit of profit is the only good.

APRIL 2006: I am in Igloolik with a Canadian film crew, and we are going with five hunters out on the land to look for polar bears. As we leave town, an old man is singing. He hangs his bare arms, palms up, out of the window of his house and tilts his face into the spring sun. "Aya, aya, aya," he chants. I'm going to spring camp at Mitliluk with Theo Ikummeq; his three nephews Harry, Joe, and Bruno; his childhood friend John Arnatsiaq; and the four-man film crew.

The camp we are headed for is a hundred miles away. It's already three in the afternoon and cold—around 20 below. Accustomed to traveling by dogsled, I ask how many days it will take to get there. Theo laughs. "We're modern hunters. Dogsleds are too slow. We'll be there tonight."

Nothing could have prepared me for the violence and deep cold of this journey on a machine. "We used to use dogsleds and wear one-piece caribou suits," Theo says. "Now we drive snowmobiles and wear white man's clothes with caribou socks." When I ask why, he looks away and shrugs.

At three in the afternoon, we head out—six snowmobiles and 14 people in all. I'm being towed behind a snowmobile on a flimsy *komatik,* a wooden sled, gulping exhaust as we fly over rough ice.

We are traveling fast, far too fast, careering over snow-covered land down onto ice. We go up and over Coxe Islands, Richards Bay, across the crooked arm of a strait, over Amherst Island, then traverse the edge of Fury and Hecla Strait until we reach the northwest corner of Baffin Island.

When dogs and dogsleds were in use here, this trip might have taken three days. You could see the shapes of the snowdrifts, feel wind changes, see animal scat and tracks, sleep in tents, tell stories, make food, sleep close together, and, while traveling, avoid the horrendous wind chill.

As the speed increases, I support myself with a straight arm. The jolts are wrenching, and at the speed of 30 miles an hour the windchill drops the temperature to minus 50. A peach moon rises over Ikiq. It is full and looks like something to eat. But we're going too fast to tell which way the wind is blowing or where we've been, and I'm at a loss to understand why anyone would choose to travel this way.

It takes nine hours to reach camp. At midnight the pink moon deepens to the color of a rose. There's a tiny cabin with a sod roof, not big enough for all 14 of us, so the drivers move their Ski-Doos in a wide circle and shine their lights on John, who has begun building an igloo.

Between blanks of night and ice light a knife flashes. Its curve-cuts come fast—*whoosh, whoosh,* and a block lifts out of the snow-bank, whole and glistening. *Whoosh, whoosh*—the knife swings again. Is it cutting darkness or ice?

The sides of each block are shaped, the edges beveled, then John sets it between other blocks, tamping it gently on the top with the handle of his knife until there's a soft thud and it shifts perfectly into place. Another rectangular block comes free, and another. As the walls go up, we chink the cracks with snow. In an hour the "boss block" is carefully set in the hole at the top and the igloo is ready.

Inside caribou skins are laid on ice benches. Our duffles are thrown in, sleeping bags unrolled. The single flame of an old primus stove is lit, giving off heat and light. "There are tidal cracks just off the point here," Theo says. "In the winter, we set nets there to catch seals. Caribou, seal, polar bear, arctic char—it's abundant here, so that's why there's a camp here. The name Mitliluk means 'sea-running char' and tells you where you are, and that you will be eating good fish here.

"A name prescribes and describes," Theo explains, taking off his winter-camouflage parka. "It tells a story about a certain kind of ice, where it's safe to hunt walrus, or it is unsafe to walk on. You need to know these things. Without those names, you won't know. You may know how to get home, but you don't know how to behave where you are."

He tells me that in a southeasterly wind, there are two kinds of snowdrifts. "One is cheeklike, the other drift is like a tongue. Then the cheeks turn into tongues, where the wind has driven through. We have words for wind and all the ways it carves snow.

"See, I'm thinking in Inuktitut. But the youngsters are thinking in English. That makes it hard for me to explain these things to them. Without language, there can be no proper transmission, there is no finding your way to a polar bear and then getting home."

I make tea for John and Theo. It's three in the morning and we're still talking. John says that snowdrifts are more accurate than wind for finding one's way. "I just drag my foot a little in the snow, and that drift will tell me which way the wind is blowing," he says. "Not so many know how to read drifts anymore. The people have lost a degree of independence that will never be recovered or know what one learns living out on the land and ice."

Theo says that there are only a few left who know the traditions. Some don't even hunt seals anymore, and it's the same with moving-ice hunting and the winter walrus hunt. "When the hunt goes, so does the reason to speak the language."

A night wind blows and the sod roof seems to shift. The low entry door made of driftwood creaks. My toes, frostbitten many times over the years, hurt now as they thaw.

"My grandmother survived on human flesh. Her husband offered himself!" Theo says and looks to see if I'm shocked, but I'm not. Starvation stories are abundant in the Arctic. He continues, "It was my father's mother. She was gone by the time I came to be. A young couple, they were headed for their autumn camp with their two children when they discovered their winter cache of food had been raided by wolves. They stopped and hunted for caribou but with no success. They jigged for fish. Nothing. Winter was coming on fast, and they knew starvation was near. Finally, they began killing their dogs for food.

"They tried to catch rabbits but couldn't find any. They ate leather thongs and straps, and later they ate their own clothes. My grandmother was only 14 years old. Her husband was older. He said to her, 'When I die, please eat me. I taste like caribou.' And finally she did. Then she ate her two children after they died.

"She was found in May. She had eaten all her clothes and had only a piece of skin to kneel on, sleep on. A couple was traveling by on their way to Pond Inlet when they heard a strange, low noise. They went towards it. They saw this woman. Her igloo had half melted and fallen down around her. When they recognized her, she said, 'I am not human. I have eaten others.'

"They gave her a little hot water, then a little boiled meat. Every hour or so they gave her a little more. When she was strong enough, they took her back to Igloolik. A tent was put up for her away from others, where she had to live for a year. Later she became a special kind of person. She married and had five children. She never let anyone in the village suffer. She shared her food with everyone so that no one should feel hunger again."

Theo sticks his head out the door. White up there, white down here. "Tomorrow won't be good weather," he says quietly.

MORNING. Snowdrifts lean against the low entryway. Theo's nephews clear it away. Theo stands outside in his insulated camouflage suit. Coffee cup in hand, he stares into the whiteout: "We could hunt today if we had to, but we'll stay here. I hope we can hunt soon. Meat is a requirement. I get heartburn if I don't eat meat!"

Back inside he and John tell stories. The two men, now in their late 50s, grew up together. John begins: "Once a polar bear grabbed me by my caribou pants. When a bear comes, two people should not stand together because he can grab both of you at the same time. You can tell by the ears which bears are aggressive and which are tame. Sometimes a bear will kill a person by chance. But if a bear doesn't want to be caught, he'll turn into something else. I've seen it happen."

Theo: "A bear might go behind some ice and come back out as a wolverine." John says he saw one turn into dark ice and another one that became a seal.

Theo's eyes shift back and forth to see if I believe. I smile and shrug. "Who's to know?" he says finally. "With the Christian beliefs we have today, these kinds of things are impossible. But before, it was believed to be real. Once you believe, then . . ." He doesn't finish his sentence.

Soft clouds lie on the horizon and the sky is clearing. We stand outside to get some fresh air. John tells me there are still powers at work here. "I saw a mermaid in one of the lakes near our summer camp." A mermaid? I ask. Was it Nuliajuk? The goddess of the sea? He shakes his head. In Rasmussen's ethnographic notes there are references to mermaids and also mermen, dwarfs, giants, shadow folk, claw trolls, mountain dwarfs, glutton spirits, the spirit of stones that could marry a woman, and the thrashing spirit that lived inside the bearded-seal whip, to name just a few. Do myths come from experience or are memories part of the collective mythology?

"From what we've been taught, there are some we can shoot and some we can't," Theo says. "A hunter has to have a feeling about the state of the animal, the accessibility of the animal. We dismiss one that is not going fast enough, or if a walrus is rushing back and forth in front of you for no reason. We don't just hunt everything that moves."

John says that the most difficult part for a full-time hunter now is getting gas for the snowmobile. "I hunt for four boys and three girls. I have to get gas from a family that works a job. Then I hunt for them too. We give each other assistance just to put meat on the table. That's why the number of hunters is going down."

Theo: "If you look at the almighty dollar, that's what's affecting all of this. The dollar has killed a lot of culture, of us being a sharing community. The dollar has had the most effect. The second thing was the missionaries. But the Bible and the Inuit ideas of sharing aren't really so different. At the time the priests came here, it didn't seem to be bad. But later, there were more serious consequences."

Like what? I ask. He glares at me for asking a stupid question. "They didn't see what we had here."

IT HAS BECOME DAY. Theo, Harry, and the film crew go out to hunt for nanoq—polar bear. The rest of us stay behind. It's minus 20 and still blowing, but we can see islands and straits, hills and flat expanses of sea ice, and beyond, Baffin Island. A young couple who arrived late are left behind to take care of camp, lest polar bears come. They use the time to teach us how to make an igloo. After several hours the blocks of hard snow we've cut are mis-shapen; they break and the "boss block" falls in. When our tiny shelter is complete at last, we chink it with snow that melts a little, then grows firm.

Is there a story about the weather going wrong? I ask Luke, the young man. He thinks for a while and says that the weather has become stormier. "It's hard to predict where the wind will come from. I used to know what the weather would be, but now it's confusing. My grandfather told us to learn the weather and know what's going to happen and what to do as a result. Now he might not know. It's all because of that hole in the ice up north in the big ocean."

During the night Luke has a fight with his girlfriend. In anger she kicks out the side of their igloo. "It's not cool to do that," one of Theo's nephews says. Ashamed, they leave before dawn. Later, John repairs the broken wall.

At two in the morning the film crew and the hunters come back with a polar bear. The animal has been skinned, the meat cut up and laid on the komatik. It will be distributed to family members back in Igloolik. The return trip takes another grueling ten hours. It's not the sub-zero weather that's bad but the speed of the snowmobile and the subsequent windchill. I sing to myself and think about Greenland dogs: how the speed of a dogsled is just right; how the dogs are deeply alert; how, when open water is near, they'll stop and crouch down suddenly. But what does a snowmobile know of danger?

One of the snowmobiles hits a piece of rough ice and flips. Harry's leg is hurt and one of the skis is broken. With their usual ingenuity, the men cobble together a new ski with bits of this and that, reveling in the opportunity to be clever. We continue on, more slowly now.

In town the bear meat is laid out on a tarp on the snowy road in front of Theo's family house. An older woman I've not seen before comes from across the street, kneels down, takes a choice hunk, and walks away. "Who was that?" I ask.

"That was my 'father,' Theo says. "See, we name our children after someone who has gone before and honor them that way. Names are gender neutral. They go according to kinship, not sex. That woman who took the meat was named after my father, so she has first choice of meat. I call her Father, never by her actual name. When a man is named after a woman, he wears girl's clothes when he's young and is given sewing needles. A woman named after a man gets a sled."

Who is being born right now? Who will be the recipient and who the one who gives the name? Brother's name to girl, girl's name to brother. "Gender doesn't matter. But who you are named after does, and what that person's life and thoughts can teach you," Theo insists. "In the old days water was thrown out the front door of the house to quench the thirst of the dead."

Later, when I ask Leah Otak about naming, she turns suddenly angry: "You'll never understand it," she says. But why wouldn't I?

EVENING. Town is quiet. I'd asked to meet with Zach Kunuk, the director of the film *Atanarjuat—The Fast Runner,* as well as the more recent film *The Journals of Knud Rasmussen,* but we learn that Zach's film assistant has been killed and his brother has just died in Iqaluit. The family is gathering for the funerals.

So I spend my days at the Oral History Lab, trying to visualize the way things looked 200 years ago, in what Margaret Mead called "a form of life in which every detail had been polished into a consistency by a thousand years of use."

At John MacDonald's house I piece together the ethnographic notes of Rasmussen and other ethnologists—Franz Boas, Diamond Jenness, Hugh Brody, and Asen Balikci—trying to re-create a whole season in my head. I imagine living under the knotted brow of Baffin Island on the moving ice of Ipik, harpooning walrus, flensing and splitting those tough skins in an igloo clouded with the smoke of walrus, whale, or seal oil and burning moss that impregnates long hair and skin clothing, as blocks of ice melt and spangle the interior with icicles.

I watch a woman pound walrus fat into glistening strips, which she fits tightly into an oblong stone hearth. At the edges she shoves pieces of oil-soaked arctic cotton, and lights it until a wide line of flame appears, "like a smile," she says.

Physical beauty and having fun are important. In earlier times women tattooed their hands, arms, and faces. "I'm talking about trying to look good," declares Apphia Agalakti Awa, an elder no longer with us. "I'm talking about the time long ago when women tried to look beautiful." The tattoos varied. "Soot would be taken from the lamp and they would run the needle through the skin. As they were pulling the thread, the thread would leave the soot behind. They would make little designs, little lines on the chin, here and there, two at a time." Other lines were put on the forehead, the eyes, and cheeks. In that way it was thought that women looked beautiful all the time. No need for makeup.

Margaret Sunaq Kipsigaq recalls: "They used to wash, from what I have heard, just using water, and their face would be cleaned with blubber, with bearded-seal blubber. Blubber makes you clean, when you are covered with oil, you get clean. That was the way they used to wash, with Kakala. It was a stove. These are stoves that are used outdoors."

Before doctors, medicines were makeshift. Apphia Awa remembered that when someone had bad diarrhea, "pond cotton was mixed with a special dirt that would get stuck in the intestines and stop the diarrhea and stomach cramps." Dog urine mixed with snow was ingested for bad colds. Bandages were made from the skin of what was described as "mushroom-type plants." A seal's gall bladder was used to kill infections in skin wounds, and fish fat was made into an ointment for burns. Women did other things besides prepare skins, tend to the sick, make clothes, gather greens, and cook. Many were shamans, and one woman, named Atuat, had her own dog team.

Less than a hundred years later, issues of survival for the people of the Arctic are of another order. They are global and seemingly irreversible. Nothing substantial has been done to cut greenhouse emissions. As carbon dioxide exceeds 380 parts per million in the atmosphere, the chances of slowing the warming trend are nil, and soon there will be no ice, only open water.

The conservative ex-prime minister of Canada, Stephen Harper, called the Kyoto Accord "a job-killing, economy-destroying accord" and "a socialist scheme to suck money out of wealth-producing nations." Now, as temperatures move up and the world economy slides down, his words seem doubly hollow. Yet the frigid winters we've been having lately have people confused. The lack of sunspot activity may be the cause. But weather is weather, with its own moods, and the upward trend in global warming is steady.

MORNING. The temperature drops to 30 below. "More like our usual winters," John MacDonald says as we walk to the grocery store. I can't help grinning. "Are you admitting that something is different, that the climate is changing?" I ask. A faint smile comes over his face. His denial seems more like a defense against the enormous grief he must be feeling as his adopted town turns nightmarish. And he's right about the ice in the narrow strait at the top of the island. The cold continental climate and the way it is protected by Baffin Island to the north helps preserve the ice. Igloolik does not face open water the way some Nunavut towns do, and the ice through the archipelago holds all spring, not subject to open water with its wind waves. But the sea ice thickness has changed and spring comes earlier, bringing birds, like robins, for which there are no names in Inuktitut.

A young Inuk on the snow-covered path yells out to John: "Hey, what happened to climate change? It's cold here!" Earlier, Carolyn complained that kids aren't wearing enough warm clothes and come into school with frostbite. "They watch southern TV and think it's only cool to wear windbreakers and tennis shoes."

Yet Nunavut residents have been noticing the changes for years. In 2001 an elder reporting from a Conference on Climate Change in Cambridge Bay said, "All our accumulated knowledge of the weather patterns are unwritten and have been passed down for generations orally. I do not speak or write English so I cannot say I know what they talk about in regards to climate change, but there are changes in our climate and the weather."

Shari Fox Gearheard, a research scientist from the University of Colorado's National Snow and Ice Data Center, lives full-time in Clyde River, on the northeast coast of Baffin Island. She has been conducting research on climate and environmental changes and documenting traditional ecological knowledge since 2000. Now fluent in Inuktitut, she and her husband ask the important questions about weather changes, wind, sea ice conditions, animal migration patterns, traditional medicines, tools for hunting, and traveling. She's working on responses to global heating and adaptation options for Nunavut communities.

The hunters in Clyde River say that in the old days there was always one long snowstorm in the winter but now there are many; that the wind packs the snow too hard for igloo building; that Ski-Doos get lost in storms; that there's more fog. They worry about sea level rise and the effect it will have on shoreline nesting areas. With coastal erosion young chicks and eggs wash away, and eventually whole villages will be lost.

Elders from Baker Lake express concern about the decreasing water supply, saying there is less water in the lake and the rivers feeding it, that arctic char cannot be found at the mouth of Prince River, that the water is not healthy, and therefore the fish are often not safe to eat.

Through Shari's persistent and energetic efforts, the elders are pooling their ecological knowledge, keeping track of weather, ice conditions, wind directions, water currents, bird and animal migration patterns, and hunters' travel routes. Using a GPS and data-gathering system, they compare historical routes, ice conditions, and temperatures to help guide today's hunters.

THE IGLOOLIK town hall is quaintly called the hamlet office—as if this were a scenic village of thatched-roofed crofts. Brian, the red-haired town manager, and his wife, Mary, a sharp-eyed young Inuit woman, recently arrived from Sanikiluaq in the Belcher Islands in Hudson Bay. Brian is wiry, full of praise for the town mayor, and eager to keep traditions alive. "I took the job because they really needed someone up here to try to straighten things out. I came because I was needed. The most traditional communities are struggling hardest. I help run the municipal government. But it really isn't a legitimate government at all."

His office is cluttered and busy. The door is always open. People come in and out and coffee is poured. He leans forward as he talks, all earnest exuberance. "If I were king, I'd use councils in the smaller communities. The elders say they have to get these kids out hunting and fishing, but the kids say they don't want that. They want to go to high school so they can go to college in Iqaluit or Ottawa. Only a few do that. There's an enormous dropout rate. The families want them home to be providers of food, to be traditional hunters, but the kids want to leave."

When the territory of Nunavut was formed, there were 25,000 people. Now there are 31,522, an average educational level of the eighth grade, and a housing shortage. "If you are 12 years old, it's hard to do your homework with 14 other people living in the house of all ages, plus visitors coming by," he says. "There's lots of stress. That's where drugs and alcohol come in. They've been here for 30 years. Total prohibition won't work. Where Mary and I lived before, in the Belcher Islands, it was dry, so an underground culture developed. Alcohol comes into town piecemeal and is consumed as soon as it gets here. You can order so many liters at a time through a committee, but they add to it by getting people who don't drink to order their quota for them. Ed DeVries, [a candidate for] the Marijuana Party . . . set up shop here for a while. Can you believe it?" He laughs. "On the other side of town the Jehovah Witness people are burning books. And in between, all sorts of things are going on. Some good, some bad."

When Brian arrived in Igloolik in 2004, there was a 2.3-million-dollar deficit on a 4.5-million-dollar budget. This had been going on for three years, but there was so much internal conflict, nothing really happened. "The Nunavut government is supposed to be monitoring these things, but they didn't get to it. No one had been doing the books. I was told that half a million dollars had gone missing.

"I don't mean to sound too critical. The Nunavut government is young. It was only established in 1999. They thought that after settling the Land Claims Agreement it would be clear sailing. Enormous expectations have been put on people in the north who have very little education or administrative experience. The pace of change is too rapid. People here only came out of a semi-nomadic life 40 or 50 years ago."

Townspeople are targets for the extractive industries, he tells me. Baffinland Iron Mines gave a presentation on its huge open pit mine at Mary River. "They talked as if it was a done deal. They never consulted the community. They came to tell the people here what they were going to do and asked only one question. It was about the shipping route through Foxe Basin. The hunters said the east side of the basin would be better because the west side is the migration corridor for the walrus. It is where they breed and calve in the spring. The mining company paid no attention. They announced a plan to ship the iron ore down the west side, through the walrus.

"During the discussion, I noted that the elders were quiet. They knew it was wrong, but they felt the younger generation just needed the jobs."

ON THE RADIO in John and Carolyn's house, a woman is giving the school lunch menu, followed by drumming and singing in Inuktitut. I'm reading a 2006 DNA report on a hybrid polar bear/grizzly shot near Sachs Harbour on Banks Island in western Nunavut. Are brown bears moving north as the climate shifts and mating with polar bears, or was it just an anomaly?

The hunters in Taloyoak on Boothia Peninsula say that polar bears in their area are leaner and more aggressive, that one "broke a cabin in town," and that they are afraid to live in their camps because there are too many hungry bears around.

World Wildlife Fund biologists claim that the 15,000 resident polar bears in Nunavut, representing five of the Arctic's subpopulations, are declining in number in west Hudson Bay, Baffin Bay, Kane Basin, Norwegian Bay, and the south Beaufort Sea. But are all populations suffering? I ask John, and if so, why?

"Ask Mitch," John says, and leads me down the path to his friend's house. Mitchell Taylor is the controversial polar bear biologist based in Igloolik. An American who became a Canadian citizen, he created the Davis Strait polar bear project to assess the health of the bear populations between Igloolik, Baffin Island, and the northwestern coast of Greenland. He's high energy, wiry, strong, and famously contrary. "For one, I don't think the climate is changing, and secondly, polar bears aren't endangered," he says straight out and looks at me for a reaction.

"I'll leave you to it," John says, grinning as he backs out the door. I look for a place to sit. Mitch takes boxes off a metal chair and pulls it up to a table. His house is nearly empty because he's leaving soon. "It's a forced retirement," he says glumly, then grabs an unopened box of cookies from his "sealift supply" and dumps them onto a plate.

He shows me a news report about the 10,000-year-old remains of a polar bear just found. "It means that polar bears have survived some huge climatic swings. Like the Medieval Warm Period a thousand years ago when it was warmer than it is now, and the Holocene Climate Optimum 5,000 to 9,000 years ago. "Bears are omnivorous, just like us," he says. "They love seal, but they can also eat ducks, seabirds, the occasional caribou or musk ox, and they can scavenge the carcasses of walrus and whales."

Only one polar bear population in western Hudson Bay has declined since the 1980s because the reproductive success of females in that area has decreased. "I guess that climate change is affecting those bears," he admits, "but really, there's no need to panic. A USGS [U.S. Geological Survey] report came out in September that claimed two-thirds of the polar bears would die off in 50 years. That's naive and presumptuous. They circulate this photograph of a skinny, bag-of-bones bear and use it as 'proof' that they're starving. Well, that was an elderly male bear, probably going off to die, not a young female, as they claimed."

He shows me records that indicate the Canadian polar bear population has increased from 12,000 to 15,000, a gain of 25 percent in the past decade. "Davis Strait is crawling with them," Mitch says. "It's not safe to camp there. They're fat, and the cubs are strong. The Davis Strait population is now around 3,000, up from 850, and that's not theory or computer modeling, that's direct observation.

"See, people who don't live here have trouble grasping just how many polar bears there are, the huge area they cover, and variability of habitat and latitude, ice movement and temperatures. Of course, climate change is having an effect on the west Hudson Bay bears—look how far south it is." He points to the map on the way. "I just don't know why people find the truth less entertaining than a good story. It's just silly to predict the demise of polar bears in 25 years based on the current media-assisted hysteria."

We munch on stale cookies and study the map. He tells me that before getting his Ph.D. in 1982 at the University of Minnesota, he served in Vietnam. The Arctic became a refuge for him. His thesis was on the distribution and abundance of polar bears in the Beaufort and Chukchi Seas.

"The first polar bear I saw was in Barrow, Alaska," Mitch says. "I'd thought nothing could live there. Then I looked out and saw this big male polar bear running through broken ice. He was pissed off at the helicopter bothering him. But that sight thrilled me to the center, like nothing else ever has."

When Mitch came to Igloolik and started the Davis Strait polar bear project, people didn't know what the populations were or where they went. He flew all over Baffin Island, Davis Strait, and northwestern Greenland and walked a good deal of that terrain. One day he ran into an old hunter who said that a polar bear had been coming from a big hole, a den, on Baffin Island. "He showed me where. That was the beginning."

They put radio collars on bears in Canada and Greenland to see how they moved. "The West Greenland current travels up and upwells at Coburg Island," he says. "The clockwise gyre takes the ice down to Bylot Island, and the polar bears stay with the ice. As the ice comes south, the bears go inland on Baffin, up to the areas where there's snow and make snow dens there. At first we estimated that the population was between 300 and 600 bears. But they were crossing to Greenland on pack ice. Our study was in the spring only, and they had already begun moving. So we came back in the fall to Cape Dyer, Bylot, Devon, Ellesmere, and the mouth of Kane Basin. The count grew to 2,100 polar bears."

They worked during the year from field camps in Davis Strait, Labrador, and Baffin Island. There were bears in camp most days. "I used to catch and tag 33 bears in one day and 846 in a season. They were all in good shape, with good cub production.

"There's lots of controversy about polar bears and climate change. They may shift their territories, but they'll be here. Maybe not in the Chukchi and Beaufort Seas, because the ice is more volatile there and the pack ice is retreating, but in Davis Strait they'll thrive. See, climate change isn't happening in the same way in every place. The Arctic is variable.

"Anyway, like John, I'm still a doubter on the subject of global warming," he says. "The Arctic Oscillation has changed polarities. So now it should start to get colder. In 20 years, you'll see. Call me. Are you ready for an ice age?" Laughter.

His mood changes. "I'm on my way out," he says, suddenly dejected. Now Lily, the young biologist, will take over. He turns on his computer. He wants to show me photographs. He clicks through image after image of the polar bears he has caught, tagged, radio collared, and talked to. "I've seen my last polar bear," he says. The room is dark. There are tears in his eyes.

We make tea and eat more stale cookies. "I've tried all my life to make things better between animals and people, knowing I was doomed. That conservation wouldn't overcome economics, that animals are never valued more than people. You have to start by understanding what we are as a critter. The only more dysfunctional system than capitalism is . . ." He doesn't finish the sentence.

I give him a copy of my book *This Cold Heaven,* about traveling with hunters by dogsled in Greenland. He thanks me, then says, "If I was writing a book, I'd go to my own window and look at the land and see what was there. Then I'd look in the mirror and acknowledge the fact of human frailty. I'd do away with the notion that animal behavior is any different from human behavior, or that it always has to be human versus animal. It shouldn't be. We're all part of the same continuum."

THE DAY IS A BOX of twilight buffeted by wind. Blowing snow is a scrim, a screen that adds light but never clears. I'm walking the coast below town. The houses are small and ragged. Men are working on Ski-Doos, hoods up, tools lying around. Snowdrifts are draped against overturned skiffs.

The first time I came to the Canadian high Arctic was in 1991, when I spent three weeks at a seal biologist's camp. The ice that May was six feet thick. By mid-June it had melted completely.

A terrible storm came one day. It blew more than 85 miles an hour for three days and three nights. Our parcol—an insulated tent used by research camps on the ice—was coming apart, and we stayed up day and night, barely managing to keep it upright.

Our food cache was buried under 15 feet of snow. Our WC tent blew away, never to be seen again. When the storm cleared, two hunters from the village of Resolute came looking to see if we had survived. They were surprised to see that we had. "You should try living in a house with real walls," one of them said, and laughed his toothless laugh at us and roared away.

At Mitch's house I meet Sonia, who is using his spare room until she can find a place of her own. She's a blue-eyed Midwesterner in her late 50s, on a Fulbright scholarship to write about life in a high Arctic town. She has a shambling, crooked gait as a result of polio, and walks with two canes. "I came here with my parents long ago. Now I'm back to stay for a year," she tells me.

Eager to be on location for Zach Kunuk's new film, *The Journals of Knud Rasmussen,* she caught rides out to camp with whomever would take her. "I like the way Zach works, making films where he lives, about the old ways and contemporary Igloolik," she says.

Mitch interjects impatiently: "But the 'old ways' represent a lifestyle no longer considered viable by most of the young people. They've been taught to expect free money, nursing care, air travel, and gas subsidies by well-meaning people. In all of Nunavut there are only 800 'elders' over 65 years old."

When she first came to Igloolik, Sonia stayed with a family. "There were two elders providing food for all—five adopted children ranging in age from four to twenty-three." She explains that two of the youngest children are actually grandchildren whom they adopted when social services removed the children from their drug- and alcohol-addicted mother. "Another six natural and adopted children live in the community, except for one son who is in jail for attempted murder. The boyfriend of the 23-year-old also lives in the house, and the couple is expecting a baby in the late spring. All are on welfare. No one has a job."

She goes into the kitchen to make more tea and tells me how exasperating it is for a writer to live in such a household: "At last count the family had 16 grandchildren. At any time of day or night one can find people crammed in the small living room, visiting, eating a meal, or watching TV, often with the sound off since many older locals don't understand English. None of the children know how to hunt, even though they come from a once prominent hunting family. Several say they don't like to camp in the summer because they prefer to have daily showers, even though water supply is an issue here. Every month the family allowance is spent on gambling.

"Crowding is one of the worst problems in Igloolik. There are 62 families on the waiting list for houses. The once yearly sealift barge failed to bring the components, so they have to wait another year. Without hunting, there's a big vacuum in people's lives, and lawlessness erupts frequently despite the strong presence of the RCMP," she says.

Mitch says he was called at his office by two high school girls. "They said, 'Thirty dollars for one, fifty dollars for two.' I said, 'No thank you' and hung up." Alcohol can be obtained legally only by applying to the community's Alcohol Education Committee, which turns down the worst offenders. Most alcohol flows into the community via bootleggers. A bottle of bootleg vodka runs about $300, a can of beer is $10 to $15, a joint is $30. Suicide rates are 11 times higher than the national average. Only one in four students graduates from high school, and teenage pregnancies are soaring.

Sonia: "I walked in on one family where the older daughter was on drugs, and she flipped and was threatening to kill her mother. Her partner comes in and there's a brawl. The mother calls the cops, so the daughter trashes the house. By the time the RCMP arrived, they could only ask the mother what she wanted done, and she said, 'Nothing. Leave us.' "

To deal with such problems, there's a justice committee of elders who handle minor infractions but not violence. They have a counseling group that meets with couples or families. They try to work things out locally, but when the binding threads have been pulled out of a society, nothing makes sense anymore.

"A tenth of the violence isn't even reported, and almost no one is prosecuted," Sonia says. "A judge came for one day in November. He'll come back in May. The attorneys meet their clients for the first time in court. Ninety percent of the clients don't show up at all.

"Once the whole community was on lockdown," she tells me. "They flew in 25 RCMP. Leah's mother said it was so ridiculous. They could have had an elder go in and solve the problem. The justice system is alien to this culture. They're used to handling things on their own. If a couple has a fight, they're separated, or the violent one is sent to another community. They want things to remain confidential.

"In the old days the elders got together and talked things out. Women weren't treated well by our standards. They were told not to upset their partners. The traditional thought was that, if there's abuse, it's the woman's fault.

"There's also racism here on both sides," Sonia says. "Most of the Qallunaat—the white people—live on the Anglican side up here where we are. The white contractors and technicians don't seem to have much interest in knowing Inuit families. John and Carolyn, of course, are the great exception. To take two such different cultural systems and blend them. Maybe it's not possible."

Mitch is restless. He gets up, paces the room, and sits down again: "I see more similarities than differences in how the older generation thinks and makes decisions. But the younger ones— well, the unemployment and suicide rates tell the story. The prognosis is not good. What we have here is aboriginal rednecks. The society is coming apart. The elders are dying, and each loss has a huge effect on the youngsters. Each generation has less and less authority. Elders are being replaced by a void. There are some good leaders stepping up, but not enough of them. It has become a culture of entitlement."

At school, he tells me, everyone passes. There are special cases for fetal alcohol syndrome kids who can't function at all, and in other classes the teachers are just trying to keep things under control. "When the kids aren't taught properly and can't read or write, they act out," Mitch says. "They think school is for losers. They stay up all night. A girl will throw over a boy for some white guy who has more skills and money. Then the Inuk boyfriend hangs himself. There's violent crime in every village—breaking and entering, stabbings. . . . We can't leave our houses untended for an hour, much less a day, without double locking them. That's the reality of Nunavut. But out on the land, it's different. The people are great. There's a welcoming community anywhere you go who do things the old way."

In 21st-century Igloolik, as in other Arctic communities, the opposing worlds of "town life" and subsistence hunting life out on the ice and land are warring forces. Is it possible to live in such an inhospitable place if you are no longer an "ice-adapted" hunter? "It's like watching a death spiral in the making," Sonia says. "Think of it: It costs $250 in gas for the snowmobile to go out and get a seal. How can you expect people to live traditionally?"

The seasonal cycle in Igloolik now is very different from 75 years ago. In the fall the annual sealift arrives, bringing a year's supply of fuel, building materials, snowmobiles, fire trucks, guns, ammunition, food, televisions, and clothing. In January, when the sun appears, there's a townwide celebration with a qulliq-lighting ceremony, a feast, games, and a talent show. There's singing and dancing—not just traditional Inuit dances but the whalers' square dances, accompanied by an accordion with Inuktitut calls.

Sonia has recently met some young hunters who go out every day and hunt for their families and are committed to living a more traditional life, drug free. "There are about a hundred of them," she tells me. "They're wonderful young men, and their wives are trying hard too. There's always hope, isn't there?"

DINNER PARTY. Carolyn has made one of her amazing patchwork meals with piquant tastes of curry, Mexican spices, homemade wine, and fresh-baked bread. She has invited Georgia, a tiny, sassy, spry woman who is nearly 80 but looks 60. She says she came to Igloolik one summer as a young woman to help the local priest and never left. She's written a lively diary of her early years here. American born, she changed her citizenship quickly to Canadian, dropping her surname in protest against Operation Surname. "Now I'm just Georgia. That seemed like enough."

She has done a little of everything in town and helps Carolyn at school, teaching reading and writing. "I live in a tiny house in the Catholic part of town. I like it there just fine," she says.

Sonia walks up the stairs, followed by Mitch. He tells the crowd he has just been relieved of his job. "Too bad," John says. "I claimed you were personally known by every polar bear between Greenland, Ellesmere Island, and Igloolik."

Mitch tells of a time out at spring camp when the aurora came all the way down to the ground. "There was no ice and the water was crackling," he says wistfully, then tells us that Igloolik is the area in Nunavut with the most biological diversity. "We have bowheads, belugas, narwhal, harp, ringed, and bearded seals, arctic char year-round, and caribou."

The subject quickly turns to money problems in town. John says that money has taken the place of food. "It is shared out amongst family members. But somehow it's not the same. It's crass. Like begging for handouts, not getting a piece of meat and cooking and eating together with the family."

Carolyn passes bowls and plates of food around the table for a second time and says, "Mothers who work for Head Start are asking for food instead of money, because the kids and men in the family just gamble it away."

John chimes in: "Gasoline here is highly subsidized. It's $1.15 per liter. People in need are given money, and their rent is reduced to $60 a month if they don't have a job. It's ordinarily $1,500, so what's the incentive to work?"

Mitch: "The purchase price for a house like mine with crummy fixtures and no insulation is $250,000. Who can afford that except a government type like me? Phone reconnections cost $250, so if anyone fails to pay their phone bill, they can't then afford to have it hooked up again."

Carolyn: "There are no jobs. And no one from the Qallunaat community can dictate what economic priorities they must hold to. Suggestions can be made, but we can't meddle in how money is spent."

John: "It's estimated that each kid in town spends $5,000 a year on soda pop at $2.50 a can. It's emblematic of the great boredom here. On the other hand, money is still used in the old way—the sharing aspect of the Inuit community."

Mitch: "The elders who are getting their compensation checks of $40,000 are preyed upon by their own families. There's pressure to share it all. The old ways of sharing don't always translate well into the 21st century."

John: "And no one is paying attention to population growth. There are 50 to 60 live births per year here. In ten years there will be another 600 people in Igloolik. It's already too crowded. No one can provide wild food for their families when everyone is putting pressure on the same resource."

Sonia: "There's some work being done on the drug and alcohol problem. In Clyde River they're teaching hip-hop. You have to be drug and alcohol free to come to the class. No more drum dancing, I guess."

Mitch pushes back from the table and gazes out the window. He looks sad and drawn. "I won't miss town, but I'll miss being out on the land. Last spring I saw a Thule site with a whalebone arch on Devon Island. It was the most beautiful house site in the world."

We move into the small living room, its bookshelves lined with tomes about the Arctic. I thumb through a book of old photographs of igloos. John says, "The hunters still build igloos when they're out, but kids kick out the walls of them if they are built too close to town—just to be mischievous."

Sonia: "Did I tell you about book burning?" I shake my head. "The Pentecostal Church that arrived here in the 1980s is having another book burning soon. They burn videos, shamans' drums, porn, and my meditation manuals. The last time, they said the smoke from burning books took on the form of *Satanisti*—Satan.

Mitch: "A deaf mute in town robbed a grave and put the corpse on a widow's doorstep.

John: "There are only two interesting things about love affairs: How they begin and how they end."

Mitch: "Like the one I had with a polar bear I'd caught. I tranquilized it and was tagging it, when it rolled over on top of me. I couldn't get out from under it. Maybe I should have stayed there."

Later in the evening we discuss the famous adaptability of the Inuit. "It is predicated on the traditional hunting life," John says. "When the bowhead whale was hunted to extinction here, people moved to other areas to access other kinds of animals. It's absurd to think that the resources right here, a day's snowmobile ride from town, can support an expanding community. And yet the ethical question arises: Do we leave people alone to live their ice age lives? Or do we make generous offers to provide for the welfare of all Canadian citizens but insist on a centralized delivery system, which, in turn, annihilates the culture?"

Sonia: "For every life saved from TB, how many suicides, stabbings, alcohol- and snowmobile-related deaths are there?

John: "As our mutual friend, the anthropologist Hugh Brody, says, it's not a matter of choosing between the traditional as opposed to the modern, but the right of a free indigenous people to choose the components of their lives."

IT'S LATE AND WE'RE STILL TALKING. Improbably, Carolyn's potted hibiscus has sent out another pink blossom as if in anticipation of spring. In a few weeks the first sun will appear. When people lived out in their camps year-round, she tells me, they celebrated the first day of the sun by blowing out the blubber lamps and relighting them. If the sun returned before the first new moon of the year, spring and summer would be warm; if it came after, it would be cool.

John says that in some Arctic encampments the sun was greeted atop igloos with a barking shout of joy, and people claimed that afterward, they could hear the sun's hiss as it quickly set again.

After dinner I walk Mitch and Sonia back to their shared house. Bands of teenagers roam the streets. Three boys hit hockey pucks back and forth on the icy streets. A young kid on a snowmobile jumps over a hummock of ice, catching air and slamming down so hard that a piece of metal flies off. He continues on, laughing.

An RCMP truck is parked at the medical clinic. "That means something bad has happened," Mitch says. Inside his house we make tea and talk until two in the morning. I walk home alone. A few groups of teenagers are standing around idly. The RCMP truck hasn't moved; the clinic is ablaze with light.

I can't sleep. The windowless room pushes in on me. I've been feeling ill and doze off but wake suddenly, gulping for air. Lucien Ukaliannuk had said, "Peculiar things happen when someone is going to pass on. As people, we make our own future. Very much so."

I'm not sure how to make sense of what I've seen here. There are stories of diminishment all over the Arctic——not just here. This is the somber tragedy of the subdued, bewildered, canny, fate-befallen Inuit people, their dazzling genius draining away with the melting ice. They have in their cultural baggage all the skills to survive, but time is not in anyone's favor these days. In Igloolik and other Arctic villages, the elders who remember life before moving to town and can teach the young ones about it are dying. Youth abounds, but their threads to the past have been cut by the sharp edge of the modern world.

MORNING IS TWILIGHT. Something has happened and I don't know what. I come up the stairs from the basement to a room stiff with grief. Outside the sky is violet, inside nothing looks human. John's and Carolyn's faces are shells. They are doing things: packing up blankets and food, *pajuktuijuq,* food gifts that will be sent to another's household.

The room seems dark, though it isn't. The human world of passion and communication has been erased. I think of my dreamed-of eye that began this chapter, roaming the room, asking to see its own death. John and Carolyn move like shadows. It is a room where too much has been seen. Earlier it was cluttered with the paraphernalia of active lives, with the circumpolar moon shining in. Now it feels like a place that can no longer hold light.

"What happened?" I ask.

John looks at me. "Leah Otak's brother was stabbed to death by his girlfriend last night. He was the last one in the family who provided food for Leah and the adopted children she is raising. That brother was the favorite child. Now she has lost her mother, both brothers, and a niece who was killed in Hall Beach."

They are pulling on anoraks and boots. "And you think our problem is climate change," John says bitterly as he goes out the door. "Be sure to keep it locked."

---

TEN-THIRTY IN THE MORNING. Pink sky but no moon, no sun. The kneading of the dough in the kitchen is vigorous and repetitive. John has returned and is making 12 loaves of bread to help Leah feed her clan. "The woman who killed Leah's brother had killed another boyfriend a few years ago," John says, shaping the loaves on a wooden board. "She did time, was released, and killed again."

Loaf pans go into the oven and others are waiting on the counter. "Leah's brother had some problems, but he'd straightened up. He was interested in theoretical physics. He studied it closely and played in a band with Zach's brother. Yes, that's the other news. Zach's brother died of brain cancer yesterday. The week before, Zach's other partner in the film company was killed violently. We'll be going to funerals," he says quietly, and for the moment his rage subsides. It's replaced by a suffocating silence as if the house had been wrapped in cotton. No Christmas tunes play. By the time John is finished baking, darkness is overtaking the town. It's 1:30 p.m.

A week earlier Leah had told me about her summer cabin at the beach, where she goes to collect her sanity. "It's amazing to take the kids there for a weekend. I don't have a lot there—no lights, only a woodstove. In spring and summer we watch the birds and have a contest to see which ones hatch their eggs first. The kids want to go there every chance they get. It calms them. We take walks at night during the full moon. Here, we live under false lights. But at camp, many still use a qulliq—a blubber lamp. They cut the walrus or seal blubber into squares and cook it to let the oil out. Then it's poured onto moss in a shallow oblong bowl and the lamp is lit. That's how we have heat and light without electricity."

At night they put nets out in the water, and in the morning they have arctic char to cook over the fire. "I would like to spend a whole summer at our old camp," Leah had said. "My mother lost a child, and we went back there about three years ago to put up a headstone. Who wants to be buried in town? We don't belong here." Now I wonder where Leah will go for solace, if there is such a thing.

At the end of the day John says, "Look, about global warming, we're just not seeing it here. But there have been climate shifts before this one. The Thule people were living here on bowhead whales, then, when the weather warmed and the ice began breaking up and shifting, the populations dispersed into smaller groups in Foxe Basin. They are an example of people adapting to climate change. It must have been catastrophic: Many humans died, others adapted. The polar bear evolved from the grizzly. Now, perhaps it's going back again, to being a brown bear."

He continues: "These are threats to the world as we know it. Maybe we won't survive. But we won't know that for a while. The more immediate threat to the Inuit is how to organize themselves around the government and globalization. What we had here was a life elegantly and brilliantly adapted to a difficult environment. In everything they expressed the genius of a place, a culture that found order and kept the barbaric at bay. Now I'm finding it difficult to see how it's all going to end."

LAST DAY. A rim of light brightens the gray horizon, then fades. Twilight goes to black. Across Davis Strait and north to Greenland there is only polar night—no light at all until the end of February. Here in Igloolik I stare out the window at bruised days.

A deceased person's soul is kept alive by a name. I ask who will get the names of the newly dead—Leah's brother and Zach Kunuk's—but no one knows. In the night, pond ice, land-fast ice, and sea ice converge under the umbral hat of a tilted sun, as the dead are absorbed into the living and the threads that bind them are carried in the names.

Before the Qallunaat came in the early 1880s, there was a dark-time ceremony called *tivajut*. In his oral history, George Kappi-aniq says, "It was held after the darkest period had climaxed and as the days started to return. They would make a big igloo. This occasion was held in celebration of the New Year that would see them to the time when they would be able to catch more game animals. They were able to determine the returning of daylight by the star structure. The stars start to move faster as the daylight starts to return because the Earth rotates and tilts. If you observe the sun as it goes down, it really goes fast. We appear to be stable but in fact we are moving."

Closed-system chaos, in the universe as well as this society. Without a ceremonial life and ritual there can be no passage from dark to light, from madness to sanity, from life to death. Without boundaries there is no societal cohesion, no self-discipline in which the good of the group remains more important than the individual. These were the ways that Inuit people managed to stave off self-interest and greed, to cope with the cold and survive.

In late autumn, in anticipation of darkness, or maybe the hunger that often came with the dark days when there was little hunting, incidents of *pibloktok*—Arctic hysteria. It broke out among humans, often women, and also dogs and was said to be the result of starvation and of mineral and vitamin D deficiencies. But the anticipation of darkness in the isolation of the far north can be profound, and the strong influence of the environment must also be factored in.

JOHN, CAROLYN, AND I eat dinner quietly and go to the living room for our evening libation. No more dinner parties for a while. Carolyn dozes on the couch. Kindness shows in her face even when she sleeps. Her days with children and mothers begin early in the morning and are strenuous. She rises and goes to bed. John and I are left to mull over global heating and cooling and the coming and going of lives.

We drink a "wee dram" of Scotch and listen raptly to a CBC radio program about collapsing stars and supernovae that experience death more than once. One star kicks off solar masses several times. More massive stars burn off carbon, and the temperature gets so high that the radiation produces both matter and antimatter. This consumes so much energy that the remaining mass begins to contract. Instability increases until finally, the star collapses at a black hole.

"It sounds like a description of Igloolik," John says solemnly. "Maybe we are dying now but will live again." Earlier, he had talked about the intricately woven web between Inuit views of the environment, cosmology, and survival. "Their understanding of nature, of being part of nature, was both transcendent and practical, and the narratives that emerged were meant to awe and instruct. Now it's all ending—the old ways and the people who knew them," John says.

And the ice, I add, finally getting in my last word about climate change. Culture is always changing, and so is climate and the extinctions it causes. But we've tampered with nature too much, this time, and it is having its way with us. It's far bigger than Igloolik.

John looks askance at me, then smiles: "I guess we're just a failed species, eh?"

Another wee dram and we listen to the end of the radio report: "After the collapse at the edge of the black hole, something else happens. There's a collision between the star shells kicked off by the pulsing star. The resulting kinetic energy produces light."

ICE ISLANDS IN WIDE LEADS. Sunrise merging with sunset and nothing between. I buckle my seat belt and click through the images on my camera. There's one of a towering thundercloud with a hole at the top, a hole like an eye in the forehead of a coming storm. When we die, can we see ourselves vanish?

The plane lifts off from Igloolik. Below, the frozen, white-blasted town appears antiseptic and harmless. In Samuel Beckett's novel *The Unnamable,* he wrote: "They must have explained to me, someone must have explained to me, what it's like, an eye, at the window, before the sea, before the sky. . . . I must have wanted it, wanted the eye, for my own."

As the plane bumps upward, I recall John's story of Sikuliarsiujittuq. It is a red star that rises on midwinter evenings over the sea ice and is named for a big man who was afraid of going out on the ice for fear that he would fall through. At last he is convinced that he should go hunting. When he asked how one should spend the night on the ice, the other hunters said it was customary to have one's hands tied up. He submitted, and later they stabbed him. He then rose up and became a star. "They love these stories about those who are harmed by others and then get otherworldly grace," John says.

Above, the stars; below, the tidal crenulations of shore-fast ice in collision with moving ice. I begin to feel the burden of this town's sorrow lift. Craning my neck, I look for Sikuliarsiujittuq, but clouds have gathered over Foxe Basin as we fly southeast toward Iqaluit. A storm is on the way.

The scene back at the airport was crowded with members of the Kunuk family. The coffin bearing Zach's brother arrived, just as the coffin bearing Leah's murdered brother was being pushed into the cargo hold. The killer, it was said, was one of the passengers on the plane.

It seems important to remember that Igloolik's problems belong to everyone. They are symptomatic of the kind of unhealthy societies that have been created everywhere—inner-city strife and rural community boredom, the cruel dispossession of colonialism, the injustice of having been betrayed, and the two kinds of imposed poverty, material and spiritual, that bring about similar results: a sense of deprivation and displacement.

The ice mirrors ruined lives, and now the ice itself is headed toward ruination. The clues to finding one's way home, so to speak, are being cruelly removed.

"We are all immigrants to the modern world," Kai Lee, a social geographer, said. But what kind of world is it? What will it be in 50 years?

The latest report from climatologist Susan Solomon at the National Oceanic and Atmospheric Administration explains that the upward trend in temperatures as a result of carbon dioxide emissions is irreversible for at least a thousand years, because the balance between heat transfer and deep-ocean mixing has been lost. The thermal expansion of seawater alone will cause 1.3 to 3.2 feet of sea-level rise, and that doesn't include melting ice sheets and glaciers, which will add much more.

OUT OF DARKNESS COMES transformation. The charnel ground is a kind of radical space, and in it exuberant, creative enterprises can suddenly flourish. They help transcend betrayal and punishment because they focus on what is, not the bitterness of the past. In Igloolik, this is happening. Young hunters are staying true to tradition and providing their families with food. The Artcirq circus troupe started as a result of a teenage suicide and is now thriving. Isuma, the film production company, has created a flood of new films by Igloolik residents, and its new Aboriginal Peoples Television Network is streaming video in Inuktitut online day and night. Zach Kunuk has just received funding for a language and culture institute that will install public access production studios in any Inuit community that wants to participate. The Women's Film Cooperative has just finished a new film called *Before Tomorrow,* which showed at the Sundance Film Festival. *Atanarjuat* and *The Journals of Knud Rasmussen* keep winning international awards.

John's oral history project is one of a kind and stands as a model for cultural preservation by indigenous peoples all over the world. It can be used locally, to inspire, to teach, to help keep a culture intact. Outsiders can read the histories to appreciate the splendor and sophistication of these circumpolar cultures.

But changes that will denigrate these cultures even further are still coming. The Northwest Passage mania has begun again. With increased summer ice melt and longer open-water seasons, world powers are salivating at the prospect of using the passage as a permanent shipping route between Europe and Asia. They've never paused to think that this traditional migration route, used by Inuit peoples coming from Asia to Alaska, Arctic Canada, and Greenland for thousands of years, will destroy what is left of the maritime ecosystem of Nunavut and Alaska including the last of the subsistence hunting villages along the way.

What have we lost, what have we gained, and what's next to go? Narwhal, whale, walrus, and seal habitat, coastal villages and seabirds, hunters and circus performers, filmmakers and translators trying to hold in their memories a spirit world that enclosed

the earthly one in ritual circuits and animal-human transformations, and a material ingenuity that allowed them to thrive in a
part of the world where most people would be dead in a day.

Out the window a wild, icy beauty unravels beneath me: frozen water, snow-covered land, rough ice and rock, raven and walrus. Some people thought that the moon spirit caused the weather
to be cold by whittling walrus tusks and strewing the shavings on
the earth like snow. How is it possible to locate tragedy in such a
fine place?

I doze a bit. I'm feeling seriously unwell—dizzy and debili-
tated as a cough that will later turn into viral pneumonia comes
from someplace deep under my ribs. At the last moment, before
leaving Igloolik, I tried to find a Kenn Borek Air plane going to
Greenland—one of my heart's homes, my refuge—but no one
was going there, not in the dark time of year.

There is no life up here without ice, dogsled, seal, or the walrus's gasping breath, the slide of the pale beluga, the diving auk,
the glittering frostfall, the polar bear, the snowdrift spirits, and
the dog-trot symphony. But I'm dreaming of another place—
Greenland—where the old way of traveling and hunting still go
on. Or will until the ice is gone.

I jolt awake. Have I been sleeping? We're flying through thick
clouds. As the plane reaches altitude, I'm hit in the eye by the sun.

# THE END OF ICE

## GREENLAND

*"Some just say the spirit world searches for us. It wants us to listen."*
—Linda Hogan, Native American writer

FEBRUARY 20, 2007. Midwinter, and Greenland's sea ice is rotting. Graying pans spin and collide with shore-fast ice. Winter sun-fires ignite anything frozen. Wind waves uncoil under sea ice, bucking and jolting until the white ice lid heaves up and shatters.

We fly up and over the thick waist of this biggest island in the world. The mist tears open, revealing an infinite horizon. The ice sheet drips like hard sauce, tonguing nunataks—mountaintops that stick out of the ice sheet. At cliff edges, ice breaks: Crevasses shatter and cascade down. The ice skin is old, stippled, threaded with dirt and turquoise. In fjords, trapped drift ice chokes inlets. Glaciers dangle from the great sheet and slide waterward, stubbing their floating toes, their long tongues snapping into extinction's silence.

Through no fault of their own, the Inuit subsistence hunters of northwestern Greenland and the animals upon which they depend for food are vanishing. Winter sea ice that was routinely 12 to 14 feet thick is now only 7 inches thick. "Eighty or ninety percent of our food comes from the ice," a hunter says.

"Without ice we are nothing. Without ice we can't travel; without ice, we starve."

Below, a stranded iceberg is decapitated. Its ribbed roof ruptures, a porthole drops, a leg of ice bends, making a knee that spins and dissolves. "Are we dying or coming into a different way of living?" an Inuit friend on the plane asks. Wind punches us down toward the Kangerlussuaq Fjord. A fringed cloud the color of spilled claret cuffs the wing. The fjord is milk. We follow it inland for 103 miles. Cerulean tarns, stippled ice, and grooved rock walls flash by as we twist through orange air. It's morning but twilight. The sun, accustomed to lying below the horizon since October, barely shows now in late February, just a few minutes before diving down.

Ahead is a blue floodlight: the terminus of the Greenland ice sheet, a remnant from an ice age that ended 10,000 years ago. Musk oxen graze a ridge. The plane stops at the glacier's snout, turns, and taxis. We left Denmark at 9:15 in the morning and flew four hours. Here, it is only 9:30 a.m. but not yet light, as if we'd been moving backward and forward simultaneously. There are "ancestor stories" that tell of time being lost, of strange beings who live at the edge of the ice cap who can make a year seem like a day. "Time is very wide here," my Greenlandic friend says. "We've now entered aboriginal time."

North. Only north. And the disburdenment that comes with it. Once we start the slow progression up the latitudinal ladder, everything else falls away. Ice is home. For the Inuit it is the white world, not the green of a savanna, that greets the eye and instructs the mind to find food and animals, make shelter, grow in happiness.

---

SEVENTEEN YEARS AGO I arrived alone in Greenland with no sure idea of where to go or how to travel. I carried under my arm two thick compendiums of ethnographic notes by Knud Rasmussen from his 1917 trip to the north of Greenland, as well as his epic *Narrative of the Fifth Thule Expedition,* describing a three-and-a-half-year dogsled trip from Greenland to Alaska.

From Rasmussen I understood that in the Arctic the path of ice was full of odd pairings—danger and plentitude, famine and beauty, humor and sorrow—and that whatever came one's way was to be met with resiliency, flexibility, modesty, self-discipline, and grace.

The rigors of this northernmost subsistence-hunting society in the world were modulated by what Rasmussen called the "wizardry" of many local shamans, who acted as intermediaries between populations of spirits, animals, and humans and whose narratives about these different beings have been the carriers of traditional ecological knowledge into the present. I learned that all skins, species, and boundaries were permeable—the polar bear could become a human, the human could be a seal. The constant hunt for food was part of a circle that, once broken, began the unraveling of the entire Arctic ecosystem.

Now frigid winter weather in Greenland is becoming an anomaly. In the International Polar Year of 2007, the ice in the Arctic Sea shrank to its summer minimum, and daily temperatures in winter were what Inuit people would describe as hot. When one hunter, Mamarut Kristiansen, experienced a temperature of 50°F, he said he hoped never to know that kind of heat again.

Winter in a polar desert should be cold and placid. Now, because of storminess, it is difficult for sea ice to form. The walls of ice and frigid air that once protected the high Arctic are disintegrating as oceans absorb more solar heat. Everything breathes:

Warm ocean water exhales, upping the air temperature; warm air comes back down as precipitation, and oceans grow even warmer.

At the top of the world clear, frigid air masses functioned as the Arctic's insulation. They protected the many expressions of deep cold: ice sheets, tundra, permafrost, snow cover, and glaciers. Now these cold treasures are being plundered by global heat. Snow and ice give way to temperate storms. Wind waves batter seasonal sea ice from beneath. Snow comes down, blanketing and incubating ice.

Little did I know when I first came to the high Arctic that the climate had been changing for decades, that global satellite monitoring of the Arctic would show that seasonal sea ice had begun its sharp decline in the 1970s. Cloud cover was decreasing, Arctic rivers were transporting larger quantities of dissolved organic carbon to the Arctic Sea.

Before the climate began to change, Greenland's indigenous culture was thriving. In the far north, to be a provider of food for dogs and families was the most honorable position in society. "Our culture was not in danger at all. It was not even threatened. We had it all here," says Jens Danielsen, one of the great hunters of Avaranasua, the farthest north district of Greenland. "We live in modern times but we keep our traditions with us. We hunt with harpoons and use cell phones; we travel on sea ice by dogsled; we made sure that snowmobiles were banned."

Now Jens, my longtime guide and friend, spends his days trying to figure out how to salvage an ice-adapted subsistence hunting society that is finding it more and more difficult to provide sufficient food. "We used to hunt from September through June on the sea ice. Now we don't know when the ice will be strong enough to hold us. We had multiyear ice that never melted. Now it is all *hikuliak,* new ice. One day it is good; the next day it will not hold our dogsleds, and on the third day it is gone completely," Jens says. "And so it will be with us."

In the days before the ice began to wane, I savored these long trips that could take two days or a month. I stopped "watching the watch." I wandered. I made new friends. To be "weathered in" or "weathered out" was a luxury, either way. I watched one season pass into another, saw ice come in September in transparent sheets that looked like water. Spring was often sunny and frigid. Seals hauled out—black commas on oceans of ice. In mid-June, the platform of ice on which polar bears, walruses, and ringed seals had flourished quickly dissolved: A whole nation vanished overnight, but only for a few months. Then the ice quickly returned.

Sun was welcomed. To greet the sun when it appeared over the horizon required pushing one's hood back and removing one's mittens. Hands were outstretched in a gesture of gratitude and acceptance. There was no thought that the ice wouldn't come back, that sun and water would be enemies. Sun and cold, animal and human life, food and shelter were bound in a deep alliance, and water went only one direction—not toward more water but toward endless square miles of ice.

Of all the nations bordering the Arctic Sea, Greenland has the northernmost continuous inhabitation. Against all odds, Greenland's subsistence Inuit hunters have maintained their traditions into the 21st century. Ironically, the west coast of Greenland is the hardest hit by the climate crisis, and as the ice goes, so go these last traditional ice age hunters.

Travel in the northern district is by dogsled only; snowmobiles have been disallowed except for emergencies. Sea ice is the highway. The hunters wear skins. Polar bear pants, fox-fur anoraks, sealskin and polar bear *kamiks* (knee-high boots) and mittens, and bird-skin underwear have been keeping these men and women warm since they began drifting across the ice of Smith Sound from Baffin Island 5,000 years ago. The sled dogs that once carried their loads across the polar north now pull their long freight sleds up and down the coast of Greenland. They are the descendants of the first dogs that accompanied Inuit people from Siberia, at least 15,000 years ago. When the ice begins to break up in spring, kayaks are lashed onto dogsleds. The hunters travel as far as they can by ice, then finish the journey on water.

In Greenland I saw how the social and cosmological landscape interacted with the physical. How landforms and oceans of ice determined when and where a glacier would move, as well as how Inuit genius would flourish living on the world's most dynamic surface—ice—and how it shaped their imaginations in the making of houses, oceangoing umiat and kayaks, masks, songs, dances, clothing, and tools. Greenlanders live with an evolving knowledge of their natural world and participate in a daily engagement with weather and the environment. Cold, storms, drift ice, fog, open water, and the constant search for food order society. Success as a hunter demands a deep intimacy with ice and fearlessness in the face of hunger.

WE ARE FLYING NORTH from Kangerlussuaq to Ilulissat. The name comes from *iluliaq,* meaning "iceberg." Perhaps the name should be changed to *auktuq,* meaning "melt." To see the coast of Greenland from the air is to observe a rapidly changing climate in action.

Ilulissat's famously productive glacier, the Jakobshavn, is 4 miles wide and 2,000 feet deep. Its movement and calving rate has doubled in the past decade, sliding 120 feet per day and discharging 11 cubic miles of ice each year.

Scientists say that all of Greenland's floating ice may soon disintegrate, as well as those glaciers that have retreated inland, where the friction of ice on rock has warmed the base of the glaciers, causing basal sliding. In addition, the basin under the heavy ice sheet is allowing ocean water in and, one scientist said, is "prying them off their beds in a runaway process of collapse."

The great floating tongue of the Jakobshavn Glacier, which lies between thousand-foot-high rock walls, has begun to shrink. Since 2000 it has retreated four miles. Floating ice acts as a buttress to the whole stacked sheet of ice. When the tongue falls apart, it "uncorks the glacier," and the ice behind it, a whole sheet of ice, grows more and more unstable.

Flying north to Qaarsut, my gut churns. As the plane follows the Viagut Strait and rounds the tip of the Nuusuaaq Peninsula, I see Uummannaq, the heart-shaped island where I lived on and off for several years. Now, as I look down, it is a place I no longer recognize. What should be ice-covered straits and fjords is all open water: dark ocean and black cliffs. In the bent arms of the fjords lie Niaqornat, Ukorsisiuut, Appat, Ikereseq—subsistence villages all without ice.

From an altitude of 20,000 feet a whole watery basin is revealed. Every island, strait, and village that I've explored and visited with Inuit friends on dogsleds and boats is dark and uncovered: To the north is Ubekendt Ejland—Unknown Island—with its one village, Illorsuit, where I spent a lonely but lovely summer, befriended by Marie Louisa, a six-year-old girl.

Below, two gulls fly under the plane's wing, their mouths open as if crying out. A layer of thin clouds slides in, mimicking ice. Surely what I'm seeing is not real. Up the Illorsuit Strait the clouds break: erasure. I see the house where I lived, but I don't see ice. Only *imeq*—open water, dark heat sink. Beyond, past Upernavik,

the northernmost settlement to be colonized by the Scandina-
vians, we fly up Melville Strait, where the 20th-century shaman
Panipaq could amass a head-high stack of fish in an hour. He later
committed suicide, perhaps because he knew his world was coming to an end.

The water is black ink strewn with bits of ice, like the awful
glitter of wrecked cars. The gulls veer off, crying. What are the
laws of sea ice in a changing climate? The wind-torn, the broken, and the breaking? Buckled heaves, wind-buffeted pancake
rounds, dendritic fractures, crumbling snouts. Sea smoke rises
from open leads as if from a fire. This ruined paradise.

We fly into a deeper shade of blue, a sky that still holds the
memory of the polar night when there was complete darkness
from October to February, when there was ice. The sun rose last
week for the first time in four months. Already, it is gone for the
day, but the air is bright. Open leads in the sea ice break into small
branches. Some describe a rough circle; others widen, demolish-
ing more ice. Greenland lost 54 cubic miles of ice in 2005, twice
as much as in the previous ten years.

Spilling down Greenland's mountains is the inland ice, Greenland's great ice sheet. Once it shone solid as a jewel along the
length of the island. Now it is sliding on its own soles, too slip-
pery with meltwater to keep standing. As Waleed Abdalati, a
NASA scientist, recently said, "The ice cap is starting to stir."

We fly past Cape York, with its stupendous bird cliffs and pere-
grine falcon colonies, and soar over the village of Savissivik, where
rock from the meteorite that struck eons ago was used by Inuit hunters for thousands of years to make harpoon points. They refer to the
rock as female, a mother that gave them strong tools. Explorer Robert Peary pilfered a large chunk of it and shipped it to the American Museum of Natural History in New York, where it still resides.

Past Thule Air Base, Pituffik, Dundas Village, once a great fox-hunting haven, now an industrialized, top-secret American base built during the Cold War.

Beneath us is Steensby Land, a rump of Greenland coastline still marked on maps as "Unexplored"—a blue bulge of ice, a white flannel slope ribbed with crevasses, a glimpse into the Arctic's inner sanctum of cobalt. We enter a rolling wave of sea smoke, but rising above it, a lingering sun shines through, while on the other side of the plane, an almost full moon grows.

Qaanaaq, population 648, appears with its rows of houses perched on a hill. There's shore-fast ice, but off its two signature islands, Kiatak and Qeqertarsuaq (Herbert Island), the ice is torn. It lies in long threads like strands of hair. The North Pole, awash in open leads and floating drift ice during summer months, is only 700 miles away. Out in the strait there's a sprawl of calved ice, blocks and bergs that look like dumped furniture. This is a Greenland I no longer recognize.

I CAN'T HELP IT. I dance for joy because I'm in my Greenland "home" again. In the one-room airport I'm greeted by Hans Jensen, who owns Qaanaaq's small hotel. All year he has been sending me sea-ice data sheets and satellite images of the coast of Qaanaaq. "The ice is starting to break all the time now," he says, loading my duffle bags in what used to be Qaanaaq's only automobile. Then I see someone whose bearlike walk gives him away. It's Jens Danielsen.

"Hello," I say. "Hello," he replies, but his eyes dart with shyness. He's tall, wears glasses, and speaks with a regal voice. He has just been elected mayor of Qaanaaq. We stand at the edge of the airport and look out at a crumpled wall of brash ice. Dogs are staked out on the ice-strewn beach. Sea smoke rises in curling waves from open water beyond.

"The current directions are changing," Jens says softly. "Last year there were icebergs from Qaanaaq all the way to Herbert Island. You know how we always hunted on good ice in May? Now it is often not possible because the heavy ice pack in the north that used to protect the ice here is gone. Now there's nothing for the new ice to hang on to.

"The big ice from the north brought walrus. Right now is the time when people went to hunt for them. When the first sun came on February 17th, we always hunted walrus at the ice edge at the very tip of Kiatak. Now we can't get out there at all."

The Qaanaaq Hotel is a five-room guesthouse in the top row of houses, with a beautiful view of the islands looking west toward Ellesmere Island. The rooms are small and simple, Danish-style. I slip into my beloved polar bear pants, tighten the suspenders, pull on the sheepskin inner boot and the sealskin kamiks, and carry my anorak and mittens to the shore, where Jens's brothers-in-law, Mamarut, Mikele, and Gedeon, are harnessing their dogs. A blue bruise hovers over open water. For months we had been planning to go out on the ice, and they'd said to come in February, when it would be cold and the ice would be good. But it's warmer here than the eastern United States was when I left, despite the 36-degree difference in latitude.

I walk the icy paths around town, past Rasmussen's house, now a small museum, past the kommune offices where Jens is in charge, and the grocery store, now ten times bigger than it was when I first came here in the 1990s. I peek into the woodworking shop; a few men are building new dogsleds. At the shore, in front of the old, one-room bachelor houses, their dogs are staked out, waiting to go to work.

Mamarut is untangling trace lines. Instead of tying the dogs two by two, Greenlanders use a fan hitch, so they fan out in front of the sled, pulling when they want, tucking under the lines, and going to the back when they need to rest.

"We should be hunting walrus now," he says ruefully. "Walrus is critical. With it we can feed our dogs and our families. But we cannot go south, we cannot go to the islands, we cannot go north," he says looking out at the sea. He's the strongest hunter of the family, a man who hunts every day. To be townbound is particularly hard for him.

"I try to get some food every day. Now I never know when I'll have anything. It could happen that soon we will have to reduce the number of dogs. Yes, some hunters are already shooting them because there is no longer enough food." He pulls his sealskin ear warmer down low. He has a moon face blotched by frostbite and nicked by scars; his dark eyes are like two smaller rounds, tiny moons in eclipse. One eye wanders. With it I imagine he's able to predict weather and the location of marine mammals under the ice.

We climb on the sled, our legs hanging over one side, ankles crossed. To the west the tail end of Herbert Island is lost in snow flurries. Close to shore the white floor is so thin it cracks as we cross. Mamarut snaps the whip above the dogs' heads. *"Huquok,"* he yells. "Go faster." The dogs zigzag through a labyrinth of brash ice, and out where the way should be hard and flat, moats of open water jiggle ice pans and a biting wind takes hold.

The ice is so bad we cannot go north. Instead, we go around the bend into the frozen fjord to a hut the brothers have skidded out onto the ice. Normally, they go far and stay out a month at a time. Now they are jigging for halibut an easy walk from town. In 1998 the spring ice between Qaanaaq and Ellesmere Island was solid. Just nine years later, good sea ice is becoming a rarity. "The ice used to be deeper than my body," Mamarut says. "Now it's only seven inches thick. Twice the fjord ice at Siorapaluk has gone out completely and refrozen. We've not seen it do that before."

Mamarut's hut is warm, the *illeq*, the sleeping platform, lined with caribou skins. A box-size generator sits on a shelf. I ask him what it's for. Smiling, he points to a pointed lightbulb, the kind used for a chandelier. A palm-in-the-hand–size radio plays Greenlandic country music.

The topo map I've carried with me since 1993 is spread on the table, and Mamarut traces the lines we've drawn on the map over the years, indicating the ice edge in spring from year to year. It is a lane of open water where walruses can be harpooned. Now the ice edge is so close to town there's almost no use marking it. "There's really no ice edge at all," Mamarut says. "It's just the end line of the screw ice."

Outside Gedeon and Mikele open two fishing holes with an iron rod and skim off the ice floating on top. Mamarut gets on his hands and knees and looks down inside. "You never know, there might be a seal swimming by," he says smiling. He's trying to be cheerful, but I know he's not. He stands up and looks around with the innocent face of a young boy. "We've only been getting halibut these days. No seals, no walrus."

The three brothers unspool their long lines down into the fjord's dark water. "They say fish is good, but we don't like it. It's not strong food. We prefer eating seal," Mamarut says. Because the sea ice is so bad, he will be going by dogsled up and over the edge of the ice cap to hunt caribou and musk oxen. It takes enormous strength and endurance to make such a trip and he's almost the only one to do it. He shows us on the map the arduous route he will take: up a long valley south of Thule Air Base, and the mountains above Bowdoin Fjord. "There were never musk oxen and caribou here," Mamarut says. "Now the herds pass behind town. The herd is increasing in size. That's because of global warming. There's more vegetation. It used to be only rock and ice. But for now, until spring, we will get fish. That's the only food we have."

The wind rises. It shears off the top of the mountains, pushing snow off the ice sheet. Last week, Gedeon, the youngest of the brothers, got caught in a terrible blizzard. He's strong, lithe, and eager—too eager to be out on unsafe ice, his quiet older brother Mikele says. "We were hunting out on the ice by Kiatak," Gedeon explains. "A storm came in fast and brought much snow. We're not used to deep snow. It's usually too cold at this time of year. The ice was thin. It's always thin now, so you just go out on it, you have to."

The wind blew hard suddenly and the ice came apart. Gedeon was on one piece of drift ice and his friend on the other. The ice started moving fast, going out beyond the islands, toward Canada. When the ice began to fall apart, Gedeon's friend cut his dogs loose and the sled went down into the water. "He was holding his dogs—there were 12 of them. Then they went down too and he was standing alone, going out to sea," Gedeon says quietly, no expression on his face.

"My ice was going fast too," he says. "Dissolving at the edges. I'd cut my dogs loose, too. They were scared. They lay down flat. More ice broke off and took my sled. I had my new cell phone," he says, suddenly smiling. "Sometimes these modern things are good. It only works if you can see the village, but I could still see the buildings at the shore, so I called my wife. Marta got help. I waited. I didn't think we would live. But a helicopter came from Thule Air Base. They got my friend first, then they hauled my dogs up, and me. Right then, the ice I'd been on cracked apart and disappeared."

Storms that bring 12 feet of snow in 12 hours are almost unknown here. Qaanaaq was buried. It was hard for Gedeon to find his way around. The female dogs with puppies suffocated because snow blew into their doghouses. The dogs staked out on the ice were half-buried, but they dug themselves out. Some of the sleds blew away and crashed on the rough ice. People were stuck inside their houses and couldn't get their doors open. Other hunters who were out on the ice found shelter in huts here and there along the mainland coast and on Herbert Island. "They were lucky to be able to get inside. Otherwise they would have been blown away."

We linger in twilight at the fishing holes. The hut is cozy, and despite the disappointment over the ice, the hunters prefer the hut to a house in town. As we prepare to leave, Gedeon notices that Qav, an old man at a camp close by, is having trouble untangling his fishing lines, and we walk across the ice to help. Qav travels to the ice on his child-size sled pulled by a single dog. "He's 93 years old but he comes out fishing every day. He lives in the Elder House and keeps everyone supplied with halibut," Mamarut tells me. Providing food and sharing it is still an essential aspect of Inuit culture, and Qav has not lost track of his obligations.

He's delighted to see us and immediately tells a story: "I used to go on long trips on the ice. Way north past Washington Land, then over to Ellesmere Island and down the coast of Baffin Land. I was gone for months and months, but I always had food. Finally, they gave me a two-way radio. I had to call on certain days, so they wouldn't come looking for me in an airplane. I don't know why they worried. I was always fine. Now it's very different from when I was a hunter. Now the ice is dangerous. I'm old, but I still come here to get halibut every day for the Elder House. Yes, I still provide the food we like to eat every day."

Returning to town, the sled bumps hard. The ice is a gong, and we are hitting it. As the dogs race ahead through winding shore ice, we pull in our arms and legs and hang on tight as we bang against pitched-up slabs and broken floors that once lay flat. An ecosystem in collapse looks like this: all rupture and rattle, all scattered fragments wrenched from their ancient foundation, all translucent bits of ice drowning. We are, each of us, a whole within a whole, but the thread that holds our disparate parts in some cockeyed union has pulled out.

We unpack the sled in the dark in a mutual state of despair. A month from now the walruses, belugas, narwhal will be moving north, and female polar bears will emerge from their snow caves with their young. Ringed seals and bearded seals will haul out on the ice. Faithful to their birthing sites, their white-fur young will stay hidden on a ledge of ice beneath the snow. Walruses will use drift ice and shore ice to make their shallow dives for shell-fish, and polar bears will ride the ice like a taxi, as spring winds push and pull the floes south and north, close to shore and away. Narwhal will lift their long, twisted teeth to test the weather and eventually swim in pods of ten or twelve to the openings of ten-mile-wide fjords, where they will mate and calve.

Mamarut, Gedeon, Mikele, and Jens will head out on dog-sleds with their kayaks tied on the sides in an ice age amphib-ious brigade. Culture is as bound up in biological diversity as biology is embedded in culture. Culture is "a product of biol-ogy," biologist E. O. Wilson has written. When we lose an eco-system we are losing our thumbprint uniqueness, our way of knowing the world and our strategies for survival. This time, not as simple as a meteorite crash that ended many lives, we are willing our own extinction.

Mamarut says we will not be able to go out hunting at all this week. "The ice is too thin," he says. He looks down the coast. "That trip we took in 2004, down to Moriusaq and Saunders Island to find walrus—we could never do that trip again. That's how bad the ice is now, and we thought it was bad then."

QAANAAQ. 1998. I've been happily stranded in town for two weeks because of weather. Today there's a whiteout. A handsome older woman named Sophie rescues me from a snowbank, invites me into her small house, lights a candle, gives me tea. She grew up in Dundas village, where Knud Rasmussen and his fellow Danish explorer Peter Freuchen had lived. In her younger days, Sophie was a shaman's apprentice, a woman who made songs and played a hand drum. She knew the stories about Sedna, the goddess of the sea, who released animals to be hunted only if humans had behaved. She knew the incest story of the sun and moon and tales of orphans who, because of their ill treatment, became wise.

She knew how to use a ptarmigan head as a weather vane: The head was cut off and the beak turned in the direction of the prevailing wind. She knew the history of her own civilization and its slow destruction by outsiders, people who did not understand the path of ice. As soon as Arctic people were moved off the ice, rock, dirt, and tundra and into houses and towns, their many gods were reduced to one, "as if they were trying to make us feel lonely in the world where we had always lived," she says.

In 1721 Hans Egede, a Lutheran from Scandinavia, arrived in West Greenland—far to the south of present-day Qaanaaq—to introduce Christianity and ban the rituals of local shamans. By the time Sophie was a young woman in the 1940s and '50s, the missionaries and teachers had penetrated the northwest coast of Greenland all the way up to Thule. "They made us stop singing, but we did it anyway. Before they came, the shamans held winter séances, even in modern times," Sophie says. "They were wrapped up tight, like mummies, and went under the ice to comb Sedna's hair. That's where the seals came from. They came from her, from way under the ice. They came up for air."

Sophie stands and sings. With bent knees, she taps the small drum made of seal intestine, her body dipping and swaying as if she were at sea, tilting from side to side and up and down in a wavelike motion. Maybe *she* is Sedna, I think, with her long hair, gray now, threaded with seals and narwhal.

She sings songs about her dead husband and his ghost. Songs about the seals and the weather that is to come. She says it will be a hot place here. Another kind of desert. She tells me she has seen ghosts walk by her little house, legless and floating, that she knew when someone was "about to be dead."

On another morning she recalls the blustery September day when the residents of Dundas were moved to Qaanaaq, after the secret Cold War treaty was made between the United States and Denmark to protect Europe and North America from the Russians. "No one had lived on this site before," she says. She wasn't allowed to sing when outsiders could hear her. "The foreign religion forbid it. Now I can sing all I want, but there's no one to hear."

We eat Danish pastries from the local bakery in the grocery store and drink coffee. I listen as she sings a song about the love of her life. She had fallen in love with a Dane from the air base, but she was not allowed to marry him. Another storm comes in and the candle flickers. "Ghosts are floating by, can you see them?" she asks. Snow-filled wind rattles the windows. She looks at me and says, "Sometimes everything is clear when there is nothing to see."

The next year when I visit Qaanaaq, Hans tells me, "Sophie's mind has reversed." The following year when I return, Sophie is dead.

THE ICE IS A LAMP. With it I'm learning to see: how new ice in the autumn can look like water; how calm water can be mistaken for ice; how in the dark time, *silanigtalersarput*—working to obtain wisdom—is a possibility. I walk the village. At this time of year it is dark or darkness going to twilight. The "ice lamp" allows Sila to be the guide. Head tilted back, I see fluted cliffs that hold ice like candles and the ice sheet breaking over the edge of the island in falling cliffs of white.

I wander around the fringes of Qaanaaq and try to see what was, what is going to be. It's afternoon, and the icebergs cast long shadows. Mist breaks over emptiness as if trying to make it into something; it rises from open water, protecting it, and causing the ice around it to decay.

Mythical giants called the Timersit lived on the ice cap, but who lives inside sea smoke? At night the cold comes on, dropping to minus 20 before the windchill. Even in summer, when the light is continuous, it is possible to tell it is night without looking. The sun cools and the world looks hollow, as if nothing remained after the ceremonial life was driven away.

"I am a collector of shadows and darkness," an old woman told Peter Freuchen and Knud Rasmussen during their Fifth Thule Expedition in the 1920s. "I keep them all locked up here in these boxes, so the world will get light again." The East Greenlanders said that on the day after the shortest day in the year, water had to be scooped from the sea into a wooden vessel and poured over

a mountaintop as quickly as possible. Doing so would make the
sun rise quickly.

I make my annual visit to Qaanaaq's small museum. Moved
here in 1997, it's the modest white house built by Rasmussen
and Freuchen as a base for their seven Arctic expeditions. The
museum's collection is a constant reminder that not much has
changed in a thousand years in the lives of these boreal hunters.
"We still do things the way our ancestors did them because we
haven't found a better way. Why change it? Some of the materials
are new, but the techniques are the same," Jens tells me. Because
they have maintained their traditional hunting culture they are
not reenacting retrieved memories of how it was, complicated by
how it is now. Blessedly, there has been a continuum.

In the glass cases, among early Dorset and Thule artifacts, is a
walrus penis bone used as a snow scraper, a washrag made from
the hide of a little auk, a gull's hide used to wash dishes, thread
made from the narwhal dorsal tendon, bearded-seal thimbles,
a blubber lamp, a bird-skin undershirt, a guillemot jacket, all
kinds of harpoons, an old sled made of narwhal tusks and whalebone, a newer one made of driftwood, and a snow knife carved
from bone.

"We didn't have very much compared to other cultures, but
we used everything we had," David Kiviok, the new curator, said.
Greenland hunters observed natural phenomena closely. Then
they turned them on their back or side to reveal more mean-
ings. Direct observation did not preclude spiritual inquiry, and a
taboo-filled society fearful of losing its social order in what now
might seem to be a Darwinian world. The imagined and the real
were not thought to be separate.

What ethnographer Wade Davis calls the ethnosphere was a
place where seasons moved humans and dogs and the animals

they hunted. March was brutally cold; late spring was glorious, with the ice edge shimmering and seals hauled out in warm sun. Summers were brief. Thirty days or so, limited to the month of July. By mid-August, winter weather began, and winters were, as they said, "most of the time." *Ukioq*—winter—was the glue that bound society together. *Hiku*—sea ice, in the northern dialect— was a dynamic habitat, a blue world of impermanence that looked solid but wasn't. It mirrored the outward calm of the hunter but lit the fluid, inward flint of the Inuit imagination.

The spirit world enclosed the human world in ritual circuits. Together, people, animals, and spirits moved seasonally, following the ice. Before calendars and watches, meetings and conferences, time was told by the arrival and departure of birds and animals. Belugas and walruses at the ice edge in March, little auks around May 10, narwhal in the fjord by June. "We are following the universe," they said. "We watch the stars. They are always moving. So is the ice. Every day our ice-world is new."

In such an environment transformations between animals and humans were understandable. Here, spirit beings were always driving into the actual. As ice shifted from hour to hour, so consciousness shifted, rendering species boundaries irrelevant. Narratives animated what, in winter, was a still place. The biological and metaphysical were understood as wholes within wholes, the one never precluding the other.

In the dark time, winter dances were accompanied by a hand drum made of seal intestine stretched on an elliptical bone frame and beaten with a walrus rib. Villagers sang, *"Aja, aja, aja."* In Greenland, unlike Arctic Alaska, where everyone composed songs, only the shamans' songs had words.

As in all Arctic villages, *angakoks*—shamans—were plentiful. Sometimes half the population of a settlement had some kind of power. An apprentice learned from an elder, being sent alone at night to a cave whose entry was then closed. In the dark, the young man or woman acquired power. Then he or she went to the edge of the ice cap, where a helping spirit—a *tornarsuk*—could be called. Such spirits took on all kinds of forms: They were shapeless or tiny or manifested as an immense bear.

"Once there were bears who could understand what we say," Sophie had told me:

Once there was a man who shot up into the sky and became a star called Nalagssartoq.

Once there was a woman who made clothes out of raven feathers.

Once there were dwarves who could kick over a whole mountain.

Once there was a woman who drank up all the streamwater and it came back out as fog.

Once there was an orphan boy who became a giant.

Once there was a shaman, Qitdlaq, who led people from the other side of the sea to Greenland. Overtaken by a storm, he drifted out to sea and came on strange people covered with feathers. When they chased him, he caused a snowstorm to come and froze his pursuers to death.

Once there was a man who lived inside the earth and was so strong he could carry a bearded seal on his back.

Once there was a giant dog who could swim out to sea and drag whales and narwhal to land, and could carry its owner and his wife on its back.

Once there was an inland dweller who was a fast runner, who caught foxes and lived near Etah.

Once there were ravens who could talk.

Once there was a bachelor who married a fox.

It was a time when animals could understand everything.

2004. Nittaalaaf. It's snowing for no reason. It's 35° below zero in mid-March. *"Huughuaq, huughuaq,"* Jens yells to his dogs. "Faster, faster." Dogs scramble and bark, a few fights break out, bearded-seal–skin whips are snapped over the dogs' backs until they pick up speed. There are eight of us, four sleds, and 58 dogs. We're off on a two-month-long hunting trip to look for walrus. The sleds tip and tilt over the rubble of pressure ice at the shore and bump down hard. A line snags, a dog is dragged, Jens leans over, snaps the line, and the dog jumps up, rejoining the others.

Behind us is Qaanaaq, with its rows of brightly painted houses. How quickly it fades behind the blowing snow. Wind drifts lie in long lines north to south. We bump over them and career between thick patches of head-high pressure ice. As we get farther out, the ice flattens. Wind wipes snow off huge plates of ice, and the dogs run fast over a cerulean mirror.

All day we travel in bitter cold. Mamarut passes us, laughing and snapping his whip. The only other sounds are the ones made by the dogs panting and the sled runners creaking over hard-packed snow. The temperature is dropping, and every leading edge of our bodies is nipped by frostbite: middle fingers, feet, nose tips, cheekbones, foreheads. We lurch through patches of jumbled ice, and though it's hard to stay on the sleds, the effort to do so keeps us warm.

March is one of the two coldest months in the year. There is light in the sky, a few hours of night, and a sun that brings little warmth. Jens holds his mittened hand against his cheek and nose, where frostbite has appeared. (It takes only 60 seconds for exposed flesh to burn at this temperature.) Behind his kind, boyish face is an elegant mind. In his community he's regarded as a natural leader with a spiritual bent: one who has been called by the polar bear spirit and who, in an earlier era, might have been an angakok.

From time to time Jens looks at me, raises his eyebrows to ask if I'm all right, laughs when I nod yes, and turns back to his beloved dogs. He is my protector, my teacher on the ice, as well as my windbreak (we sit sideways on the sled, legs dangling, ankles crossed), and I'm grateful to him.

Twice we stop to melt ice for water, make tea, eat cookies. While we try to warm ourselves, the dogs roll in the snow to cool down. "Man and dogs go together here," Jens says. "It's a good combination. We have great respect and affection for each other. They aren't pets; they're half wild; maybe we are too!" The dogs have to be a little hungry to keep working for them, and they have to be hungry to keep going out with them. "They need us and we need them. We belong to each other," he says.

We cross the mouth of the fjord, where, in summer, the long-toothed narwhal breed and calve, then follow the coast of Steensby Land, named for a Danish ethnographer. No food up there. Why go? The hunters ask.

Looking west toward Canada's Ellesmere Island, the horizon is no longer a thin line of light but a feathery gray spray of mist—water sky, denoting open water where it should be frozen at this time of year. Little do we know that the sea ice of the entire Arctic is in decline, that abrupt climate change is coming on much faster than anyone expected.

As we bump along, I try to ask why there is open water so early in the spring and how ice can break up in these temperatures, but it's too cold to talk. We pull the hoods of our fox-fur anoraks over our faces and continue on.

Sun is low in the sky. Light shoots up in the four cardinal directions. At the end of the day we make camp on a rocky beach at the edge of the shore-fast ice. It's 11 in the evening and the light is fading. But the hunters are in high spirits. The farther from town they get, the happier they are. "Now we are entering some nice country," Mamarut says, smiling, meaning the ice edge and walrus. Dogs are unhitched and retied by cutting notches in the ice and threading the end of the harness line through it. Sleds are unloaded and pushed together. Two small canvas tents are pitched over the sleds. Harpoon shafts are the tent stakes; the sleds are our beds.

We lay caribou skins down, then our sleeping bags. The floor is ice. A line running the length of the ridgepole is hung with sealskin mittens, kamiks, hats, a loaf of bread, a hunk of cheese, camera batteries, and my pens, because even they need thawing. An old Primus stove is lit. In a battered pot a spangled piece of glacier ice snaps and pops and slowly turns to water.

After dinner the men prepare to hunt walrus. Gear is laid out—harpoon lines and rifles, block and tackle; knives and harpoon points are sharpened. "Yes, yes, yes," Mamarut chants. *"Aurrit, aurrit, aurrit."* Walrus. "There are many out there." Tonight, we are going to the *hiku hinaa*—the ice edge.

We walk single file for an hour. "The walrus are very alert," Mamarut warns. "They can hear us moving over the ice, so we must make it sound like we are just one hunter, moving as though we were one man."

The floor of the world groans as we walk; the ice age procession is solemn. A red sun hangs just above the horizon as if waiting. The temperature has dropped. Now it's 40 below. Thin ice undulates like rubber. We move in flickering frostfall. "It's so beautiful. I could walk all night," Gedeon whispers. The ice shifts as the tide goes out. "Now the ice we were walking on will become unsafe," he says.

The men climb a stranded iceberg and use it as a lookout. No walrus in sight. Far out gray mist unfurls from an open sea and folds back down around a warehouse-size iceberg. Venus shines in a charcoal sky. *"Issiktuq,"* Jens whispers, rubbing his arms. "It's cold!" No walrus pass by.

After a long wait, we walk the two miles back to camp. "The walrus must have heard us coming, but they'll return," Mamarut says. "There should be walrus all the way down the coast to Moriusaq and out to Saunders Island." He shrugs nonchalantly and smiles. "Now life begins to get good," he says with a deep grunt.

The dogs howl a welcoming chorus as a frigid wind lifts the hair on their shoulders. Lanterns are lit. There are four hours of darkness. A late snack is eaten—seal jerky, cookies, and tea. "We'll go back tomorrow," Jens says in his deep, purring voice. All eight of us lie nose to nose on caribou skins. The hissing Primus stove keeps us warm.

Before dawn Mamarut is gone. We learn later that he's gone to the ice edge alone and has harpooned a walrus. "He is always like that," Jens explains. "He is always going out alone on the ice. He can't stay still. He is hard to keep up with."

Mamarut returns rosy cheeked and exuberant. "The walruses were close to the ice edge. They were swimming up and down. When they went underwater, I moved in closer; when they were up, I stayed still. The next time they came up, I harpooned one. The ice was bad but it held me! [laughter] Don't worry, there are lots of walrus. Narwhal too. They'll go away for a while, but they'll be back in a couple of hours, and then we'll have enough food for all the dogs and us too."

When there is light in the sky, Jens harnesses one sled while Mamarut sharpens his flensing knives. "I'm happy that the dogs will be getting good food. They'll have lots of energy for traveling," he says. We follow the sled, moving quickly in the direction of the dead walrus. It's important to go back for it before a polar bear comes.

Ahead gray mist curls up like smoke and something moves, not a bear but a slender channel of open water ablaze in sun. "Miteq!" Jens whispers, pointing to an eider duck flying by, then two arctic terns. We approach the ice edge with a keen eye for polar bears. We step carefully: Thin ice bends under our feet, and beyond, water churns.

Mamarut motions for us to stop. There's the sound of gulping, sloshing, thrashing, blowing. A pod of beluga whales swim by, their ivory backs flashing in sunlight. We stare at them, dazed. Winter was ice tight. Now it has opened, giving life. The chilled water is oxygen rich, glutted with fish and plankton. If winter was otherworldly, so is spring. Another eider duck flies down the lead. Beyond is a delicate ladder of ice laid down flat; its rungs are blue. Rotting ice pans heave out of their turquoise moats, and ice pulls apart into long strands.

More churning. "Aurrit! Walrus coming!" Mamarut says in a loud whisper. They bob up and down, almost comically, gulping and splashing. But before anyone can throw a harpoon, they dive and swim away. It's not known exactly how far south or west the Atlantic walrus off Qaanaaq spend their winters. Some go to the south coast of Baffin Island. But in the spring, they follow the food sources north, and the breaking up of the ice.

No time is wasted. Mamarut takes out his flensing knives and stands before the one dead walrus. He looks at it with admiration: "The mind of the walrus is wild and aggressive," he says. "He is always defending his territory. Not long ago we hunted walrus from kayaks, but they attacked our boats and killed many of us. You can walk right up to them when they are resting and they can be gentle. But if threatened, they are fast."

They winch the animal from the hole in the ice, its whiskers silvered. Tusks and head appear, then the bulbous body shining in cold sunlight. A mature male walrus can weigh 2,500 pounds and live to be 35 years old. With swift, sure strokes the men cut the walrus open. Heart and liver are laid out on the ice, steaming, and a tangle of guts flow from the abdomen. Jens cuts the intestines in long lengths and feeds them to the dogs. Flippers are cut off. They look like hands. Long lengths of skin with an inch of blubber are peeled back. This is mataaq, eaten by everyone for the essential vitamins and minerals in a place where no vegetables or fruits grow.

Meat is stacked on the sled. When flensing is finished, Mamarut leaves a pile of meaty ribs and intestines on the ice. "For nanoq," he says. "For the polar bears."

On the way back to the tents, the mood is relaxed and happy. Mamarut's wild eye seems like a link to the wild animals around him. He talks about what he has learned from polar bears, how, when he fell through the ice once, he put his left arm and left leg up on the edge, using his free hand like a bear paw moving in the water and lifted himself sideways up onto the ice. "We are always learning from the polar bear. He is good for us to see all the time. The bear is his own weapon. He doesn't need a gun or a harpoon like we do. He can move on water or on ice equally and can hunt anything. He is worth our admiration. Without knowing his ways, I would have died many times."

I look back at the heap of innards steaming on the ice. Already the ravens have discovered it and are diving down to the feast. The sharing of food between humans and animals is a practical and moral necessity. "Sometimes things go against us and we don't get anything to eat," Mamarut says. "Our lives are based on how nature gives us animals. And we give food back to them," he says, speaking softly and slowly, as if looking back in time, remembering incidents of hunger in his village.

ONCE THERE was a great famine in east Greenland, when two winters followed, one upon the other, without a summer in between. Huge blocks of ice began to shoot up out of the sea, and the bottom of the ocean seemed to be covered with ice. At the end of the first winter there were no living things. The sea was ice covered all summer. The second winter people consumed their skin clothing and kayak-skin coverings. Corpses were cut up and devoured. Parents ate children and children ate parents. Then they began murdering one another for food. After eating human liver, they went mad and their hair fell out. But a few did survive. Summer came again, and all who came after are descended from the time of winter and cannibalism.

Back at camp, Jens makes a slit in the walrus stomach and pokes around in the brown juices with his knife. He spears a scallop and offers it to me. "You first," I say. He pops it in his mouth. *"Ummmm. Mamatuk,"* he says. "Good."

Sun shines through frostfall, and the hairs in my nose freeze instantly. Our thermometer reads 44 below. We've been thrown into a hall of light that no longer confers warmth. This is a cold sun, so cold it might not be a sun at all. Mornings, we wash ourselves with snow because it melts from body heat, and afterward we dry ourselves with snow because it absorbs moisture.

"In the last two or three years there were big storms and high waves in November," Jens says. "That was new. We never had storms like that break up the ice before. The ice refreezes, then we have screw ice. It's hard on hunting. There are now so few days of getting food. This year the ice didn't come until December, later than we've ever known." A thin cloud slides over the sun, making the day suddenly colder. Jens is pensive, sitting on an upturned fuel can looking out to sea. "Our lives are based on ice," he says. "Without it we can't live, we can't eat. Ten years ago the ice was six feet thick. When the ice was thick, nothing bothered us when it was cold like this. Now the ice is so thin, just a little wind wave breaks it. It has been like this for the last three years."

We sleep in white tents at the shore. When we wake, it's hard to know if it is morning or evening. "We are lost," I say, but it doesn't matter. There are larger powers at work here: Sila rules.

On the vernal equinox a front pushes in fast. "We have to go now before the storm hits us," the hunters say. Sleds are loaded hastily as wind gusts increase. The men kneel on flapping caribou skins and pull the lash ropes tight. The dogs are wild eyed. The moment the lines are hooked to the sleds, they roar off. The hunters make flying leaps toward the moving sleds, just barely making it aboard.

For days we travel south along the mountainous coast. Out on the frozen sea the world is made of wind-driven drifts and upended, see-through pieces of ice. Ocean currents have squeezed ice into a chaotic labyrinth. At an impasse the dogs moan and cry. Jens stands up on the sled, surveying the scene, then lifts and turns the front sled runners. We bump through a narrow passage, pulling our legs up to our chests to squeeze through.

A hard wind makes it too cold to stop for tea. Behind us is a bank of clouds layered black and white. Ice fog and blowing snow shroud the horizon. A single pointed mountain sticks up through the fog. *"Amaumak,"* Mamarut yells over to Jens's sled, making a cupping gesture to indicate a breast because that's what the word means. We stop and hack off a piece of glacier ice from a stranded floe, then continue on. Jens points to something buried in snow. The dogs clamber up a steep hill to it. Then I see: a tiny hut covered in snow, with only a bit of roof and part of one window visible.

We slide off the high loads on the sleds, and for a moment the wind stops. The only thing alive is ice, moaning and cracking in tympanic reports. Cold is the thing. It is a kind of night, a wall that isolates us from the rest of the living world. Two ravens tumble end over end as if their mock fall to earth might shatter the clouds and let in more sun. We check a thermometer: 59 degrees below zero. "Issiktuq"—cold—Jens says, laughing as the wind kicks up again.

The entry to the hut is dug out, dogs are unharnessed, and lash ropes untied. My fingers don't work. I help unload sleds using my wrists. Always observant, Jens quietly leads me into the hut, lights a Primus stove for heat, and gently holds my hands. "Frostbite comes quickly, and you don't know it. It's like a bad dream, a ghost putting its hands on you," he says. Windows rattle. A hunk of walrus is hung from the ceiling. When it begins thawing, large pieces are cut off and fed to the dogs.

The dogs always come first, and the humans eat what's left. Without dogs, it would be impossible to get home. A battered tin pot is put on the single burner and stuffed with pieces of ice. They crack and pop and melt. In the Arctic we are always thirsty—it's too cold to carry water with us, and living on the frozen sea, there is little potable water. We make herb tea quickly, then Gedeon puts meat in the remaining water to boil. Jens and Mamarut mend broken harpoon shafts. Dinner is walrus-heart soup.

Each hunter has his specialty. Jens's is dogs. When most of the dogs in Qaanaaq died from distemper in 1988, he was selected to choose and buy 250 young dogs from around Greenland and bring them back to town. "Dogs and humans have been together for at least 20,000 years," he says. "Dogs came with Arctic people from Siberia and were first used here by the Saqqaq culture to carry loads. When the Thule people came, around A.D. 900, the dogs were used to pull sleds, as they do today. We haven't advanced much, have we?" Laughter.

Our stomachs are full and heat spreads throughout the tiny cabin. Stories are told. Mamarut tells of being lost with his wife, Tecummeq, who is a great-granddaughter of Robert Peary. "We were out on the ice near Moriusaq and there was a whiteout. We were lost. I kept looking at my dogs to see which way they were looking. I let them go where they wanted. They can smell danger—open water or a polar bear—so it was best to let them find their way. They headed out, and then I saw there was an opening by an iceberg that I recognized. I knew where I was then. The dogs took me to an iceberg that I knew so I could find my way home."

We spend three days waiting out weather in a 14-foot-by-16-foot hut, all eight of us squeezed in tight. Aleqa Hammond, who has joined us as a translator, builds a "women's igloo"—an ice wall for bathroom privacy on one side of the hut, with a niche to hold our mittens for quick retrieval, since exposure of the skin for more than a minute results in frostbite.

We pass tins of salve and a mirror to doctor frostbitten noses and cheeks. With tweezers Jens plucks the sparse hairs from his face because, he tells me, they gather frost. Mamarut and his younger brother Gedeon sharpen knives. We eat mataaq, dipping the knives in salt first, then scoring the fat just under the walrus skin. This is the source of vitamin C and all other minerals necessary to sustain human life. "It is our orange and lemon," Gedeon says with a sly smile.

Arctic-hare socks and sealskin kamiks are dried, boiled walrus, which tastes like salty pot roast, is eaten. The men beg Aleqa to tell ghost stories and she does until the sound of snoring begins, then we sleep.

Morning. Windy and minus 50 degrees plus windchill. Mamarut peers out the one window at Breast Mountain: "Only the nipple shows," he says smiling. One of the great hunters of Qaanaaq, Mamarut is always giving credit to his teacher: the polar bear. "A man can't walk on thin ice, but a polar bear can. We learned how to cross thin ice from nanoq. They lie flat out on the ice to distribute their weight. That's what we do too."

In the evening the dogs are fed. The hut and sleds are blasted white. Jens feeds his dogs first, then Mamarut, then Gedeon, then Tupiassi—the order tacitly reflecting the respect assigned to the hunters. "Our dogs are like their owners," Jens says. "They love to eat. They are like running stomachs!"

Mamarut: "I think I just heard my lead dog say thank you."

With close to 60 dogs here, eight of us, and the extreme cold, almost a third of the walrus has been consumed. We'll soon run out of food. It is the first day of spring.

MARCH 22. Warmer today. Up to minus 35 degrees. The sky is blank and the sun is an ashen orb shrouded by ice fog. Blowing snow stacks up on the dogs and sleds, half-burying them. The hunters turn their sleds upside down and work on the runners. Rough ice and hikuliaq, first-year ice that is salty, is hard on everything—on the soles of kamiks as well as sled runners.

Later, Mamarut repairs his kamiks with a thick steel needle, narwhal-sinew thread, and a thimble made of sealskin. We eat a late night snack of what Aleqa calls swim feet—walrus flippers that are slightly gelatinous and taste of mushrooms and red meat—then the room goes silent. Jens sits at the edge of the sleeping platform and the others turn to him in a quiet reverence. He begins a story, a special one about the polar bear spirit:

"When I was a boy, my father went out from Moriusaq and saw the track of a polar bear. He followed it and finally got very close to the bear. His chance to shoot came, but just then the polar bear turned and looked at him. The bear had a human face and was smiling at my father and saying, 'Take me, I'm yours.' The dogs were scared and ran away. My father stood there. He couldn't shoot. He let the bear go. If a person has special talents, animals will come and ask to be your spiritual helper. You are only asked once by nanoq. But my father denied him and ran away. He had his chance to get the powers. After that, I was afraid of having to meet that kind of polar bear."

March 23. The Breast casts a long shadow over the frozen sea, and a city of icebergs carries morning light on its shoulders. The ice was pink; now it is dull white, an Arctic sponge sopping up walrus blood. The plan is to go south to the village of Moriusaq, then out to the ice edge at "Walrus El Dorado"—two tiny islands near Saunders Island—because with this extreme cold, they say, there will be a solid ice edge out there.

We wait in the hut until the wind dies down. The hunters line up on the illeq and listen reverently as Jens continues his polar bear story:

"One time when we were hunting at Walrus El Dorado, I went up on the land to look out for the ice edge. Suddenly, I began to feel as if there was a polar bear nearby. I had no gun and the dogs were at the bottom of the hill. I could hear the bear breathing. It was very close, so close I thought I wouldn't be able to get away from him. I ran down the hill to the dogs and could hear steps behind me. When I got to the dogs, they were acting wild, so I cut them loose and they went running up the hill to find the bear. We got our guns and followed them. We searched, but there were no tracks. Then, suddenly, the bear appeared. It had a human face like the one my father saw. I didn't know what to do. Then, like my father, I said 'No' to it. I had to. I didn't know what would become of me if I had gone with it, and I had a family to take care of. I was afraid it would kill me. Now, especially when the weather is changing, I sometimes feel it come near."

The spirit world of the Greenlanders was sent underground by Scandinavian missionaries, who arrived in west Greenland in 1721. But if the polar bear spirit is still strong, is it searching for us? I ask Jens. He nods yes. Would the world be different if we had listened?

All is ready. Sleds are loaded, whips unfurled, lash lines pulled tight. As soon as the dogs are hitched up, we sit quickly and career off the hill through snow-dusted cubes of turquoise—pressure ice squeezed by wind waves at the shore—and glide out onto the Inuit highway, made of sea ice and breathing holes.

Our single track across the ice up to the hut has been demolished by drifting snow. Now the track is a broken wave of rubble. We pick our way through the ruined ice south toward Moriusaq, population 21, where Jens and Mamarut were born and grew up and where Mamarut's wife, Tecummeq, teaches school.

To be traveling again feels good, no matter how harsh the conditions. Just before we bounce up and over a huge piece of ice, Jens clamps his leg down over mine as I grab for his shoulder. The sled tilts almost vertically, and we are laughing. I look back. Mamarut has fallen off his sled. He runs behind it, then alongside, flinging himself on, belly down.

Out on the frozen expanse the dogs run fast. They have walrus in their bellies, and they are wild. Wind drills frigid air into our flesh, burning our faces in a long strip, but I'm warm inside my sealskin mittens, polar bear pants, and kamiks. Without them, we couldn't survive.

As we reach Moriusaq, another storm comes. The villagers emerge to greet us and help unload. In Tecummeq's spacious house, water has been heated for Aleqa and me to bathe. Tecummeq's school has two students. Not long ago she says she didn't need a job because she went hunting with Mamarut. "I was always with him on polar bear and walrus hunts. That's where I'm happy, that's where I belong. I would like to quit this job and go with him but we need some money. My husband cannot make money from the extra skins because the world market won't allow us to sell them. To live in town is like prison."

We wait out the storm. For dinner, polar bear meat is served, and there is a quiet reverence as it is eaten. The men go outside and build a small igloo for fun. Mamarut trains Tupiassi's dogs to behave in front of a polar bear. Tecummeq and Aleqa make sealskin purses. I wash my long hair and braid it again.

The temperature has risen abruptly. When we look out the window we see a *puikkarneq*—a mirage that blunt-cuts the top of the icebergs. Two hunters, just back from Saunders Island, report that the ice is breaking up rapidly, that they just barely made it home, so the much anticipated walrus hunt is called off.

A meeting is held to decide what to do. "If we go out now and the ice breaks up, we might be able to ride a piece of drift ice until the winds shifts and it drifts back, but if it doesn't, then we're in trouble," Mamarut says. It's decided to go north again—a one-week trip to Siorapaluk or out to the island of Kiatak. We're trying to retrace our steps in search of better ice.

SUN DOG. Water glitter. Sun-glazed icebergs. The dogs are running fast. Trace lines catch on rough ice. Dogs are dragged, lines are snapped, dogs get on their feet in the chaos and keep trotting. We thread our way through open leads. *Hikuliak,* new ice, is all broken, and blown snow is overlapping like fish scales. Bits of ice are strewn about like cut crystals. The ice along the coast on which we traveled three days ago is gone.

An island of gray mist hangs in the air. Open water extends all the way to Ellesmere Island. We spend the night in the hut next to Breast Mountain again. In the morning we'll have to go up and over a part of the ice cap to get north up the coast. For the first time the hunters realize that the ice conditions we're experiencing aren't anomalous but have become the norm, yet they can't yet conceive of a world without ice.

"These last couple of years we have been wondering why the pack ice from Canada-side didn't come," Jens says. "It always came every summer, but the last two years it hasn't been there. We wonder what has happened to our ice. Even the ocean current has changed—it's not just that the ice isn't there, but the sea is acting differently. If there is no ice, that will be very hard for us. Even in the summer it has changed. The sea is not calm and we have lots of rain. There are rivers coming down the mountains and through our town. We never really had rain."

Another night at the hut and the wind howls. Mamarut makes a new whip from a bearded-seal hide. It is a long, thin strip, thinner at the end. "I want to be tough going up the inland ice," he says, laughing. In the afternoon we load up and cross the narrow fjord behind the hut. The Greenland ice sheet is before us. The hunters pick a smooth route and we start climbing.

Partway up the ice cliff lobes of gray-blue ice bulge out, cracked and mortared with snow. Deep drifts lie sideways like collapsed sails. Crevasses gape. At one point I jump off the sled to push from behind, but Jens yells at me angrily to get back on. Near one summit we enter a no-life zone, a soundless verticality—not quietude, but noiseless noise, a frigid din.

This is the mother ice, the center of emptiness, the umbilicus where Sila was born. It is a remnant of the last ice age, which ended 11,500 years ago and still covers 82 percent of the island. The ice slides off the mountainous coast like white frosting, feeding hundreds of outlet glaciers whose long tongues float out on Davis Strait, Smith Sound, and the Arctic Sea. From the sea ice below I've heard dry summit music: wind blowing snow across ice.

I once thought of the ice cap as an unblinking jewel light, unscathed and invulnerable. Now the glaciers are wounds with ablating flanks, broken knees, and long tongues. If the entire ice cap melts, it will have added 23 feet of fresh water to the ocean systems; it is melting at almost 8,500 cubic feet per year. If the glaciers retreat far enough, seawater intrusions will pry off the ice of its bed and the whole thing will collapse—a huge watery wound mixing with stinging salt.

We bump over translucent patches of ice that are crosshatched and grooved with dirt. Crevasses are sun-combed. One gaping ravine runs to the left of the sled, and at one point we slide perilously close. The dogs struggle: Our sled weighs between 800 and 1,000 pounds. The dogs use their claws like crampons to gain altitude.

Up and up we go, one false summit giving way to another, the dogs entering a vertical valley where long, elegant shadows cross our path, turning the world blue. Ahead an extruded hunk of ice pushes out of a wind-blasted drift, like a walrus breaking though ice, Jens says. On the other side, a series of ice ridges fall away. At the top we cruise through a snow-covered bowl, fly over a cornice, and "catch air," thumping down hard on the other side. I look for the Timersit, the giants who live on the ice cap and eagerly devour humans, but see none.

We think of the inland ice as unpopulated and lifeless, but it is full of life. Bacteria and snow algae—*Chlamydomonas nivalis*— color the edge of the ice cap and its glaciers red. Small indentations called cryoconite holes dot the surface. When cosmic dust falls on the ice cap, it absorbs heat and bores holes into the ice. Tiny aquaria form in these hollows: A miniature ecosystem arises deep inside, where iron bacteria, blue-green algae, and green algae feed nematodes, rotifers, and *Diphascon recameri* (water bears).

The ice is alive and we are dead, or are we just lost? The snowdrift's spirit laughs at us as it brings on stronger and stronger storms. Behind a torn finger of glacier we rest a moment. The hunters lay loops of soft rope under the sled runners for brakes. We take off, following a tortuous streambed down.

Jens navigates with his voice, his whip, and his prodigious weight, using his heel as a rudder. We've shed our anoraks in order to be lithe. Balance is essential on such a ride to keep from getting broken legs and smashed heads.

Jens is on one knee and drags his free foot in the snow to brake and steer. I grab caribou skins and lash ropes and lean into the

corners as if on a bike as we accelerate. The sled tips onto one runner, bangs down, then tips the other way. Finally we stop. Before
us is the sea that divides Greenland from Ellesmere Island. The
sun is going down and the horizon is golden. There is open water
as far as we can see and broken ice at the shore. The tide is coming
in, and huge plates of ice break apart and jostle restlessly. "Impass-
able," the hunters say, talking among themselves.

There are hand gestures cutting the air. Then we get going again,
following an ice foot inward, up a fjord. We have nothing to feed
the dogs, and they've worked hard. Ahead there's a hut. Jens taps
me on the shoulder and points: There's a drying rack with meat
on top—extra food the last hunter left for those coming after. Jens
smiles. Tonight we will eat soup, but the dogs will get seal.


ISSANGIAQ—SPRING WIND. For the first time it's above zero. We
cross Ikersuaq—Big Opening—at the mouth of the fjord and
travel into a world of ruined ice. Once sea ice was leaf and lid,
flattening the ocean's torment. Now it is frozen chaos, and by
comparison, the sea beneath seems tame.

The hunters pick a route that will take us to the outermost
island of Kiatak. They say there is always good ice between the
islands at this time of year. We pass a single polar bear track three
times bigger than my hand. The men stop to inspect it. The track
is fresh and going in the direction we are going. No one says anything. Quietly we move on.

Behind us mist makes runnels down the face of the glacier
with sun stripes peeking through. The air is humid and feels cold.
Kneeling on the moving sled, Jens pulls on his fox-fur anorak. We
rumble across broken surface and tipsy pancake ice in turquoise
moats and lanes of gray slush ice.

What had been mist is now flying snow and pale light erasing the place ahead that separates sky and ice. The hunters never imagined that this whole month would be spent looking for ice good enough to travel on, that there would be no way to hunt for food. In the old days, before there was a store in town, people and dogs would already be going hungry. In the Arctic, bad weather and bad ice can make things go wrong very quickly.

We travel for hours. Once Greenland was an icy paradise; now this last remnant of an ice age culture is wearing thin. "A few years ago it was even worse. A lot of people were suffering from not having enough to eat. We can only wonder why it is like this, why we can't change it," Jens says. Climatologists are already saying that if emissions aren't cut within ten years, Arctic sea ice may never return. To read about it is one thing; to feel the ice go out from under your feet and to go hungry is quite another.

Gedeon gets two seals in a net he set the previous November. They are flensed, the blubber stripped away, the meat cut up and fed to the dogs. We eat what's left. The next day we follow an ice foot around the northeast side of Kiatak. A third of the way up along the island, the ice foot ends abruptly and we are forced to jump down a 14-foot-high cliff. Below what should be solid ice is a gray, watery plain of rotting ice.

The dogs are scared. They can smell open water and know the ice is dangerous. The hunters unhook the dogs and push the heavily packed sleds over the precipice. They bounce and slide though slush. Then it's the dogs' turn, but they scramble backward, digging in their heels. Gedeon pulls at the trace lines. Finally, his dogs go over, leaping onto unsteady ice. The others follow. Gedeon and Jens jump. Mamarut ties a rope under my arms and lowers me down, because the ice at the bottom is jagged and he's afraid I'll break my leg.

Down on the ice we hurry. The dogs are hooked up, sleds are righted, whips are brandished. The dogs move fast despite their fear. Some broken ice is knife sharp, some is gray slush. Ahead, pancake ice tips and twirls as the dogs leap fearfully over moats of open water. The hunters yell commands. The urgency in their voices tells their dogs that this is serious and they must behave.

Crossing the ice to Kiatak Island a storm lowers down on us. What had been flowing mist is now flying snow. We camp at the nearest hut as the blizzard comes on. Jens says the ice we just crossed is called *tinumisaartoq*, meaning, "ice getting a hump in its back." It's 10 p.m., and Mamarut's lead dog is dying. I ask why. No one knows, but Gedeon says, "They die very quickly here." They've all been vaccinated for distemper, but sometimes the vaccine is old. "My dog had diarrhea today and threw up, but still he screwed the female in heat," Mamarut says. "He's trying to make a new lead dog for me."

Mamarut has tied the sick dog away from the others, next to a frozen dog carcass. Sadness fills the hut. There is no joking. Two small lanterns are lit, and on the wall is a cross. It's said a ghost lives in here.

Last year Gedeon had 16 dogs. Nine died. Now everyone is worried about the disease spreading, but there's nothing to be done. After dinner we get in our sleeping bags, lying nose to nose on the illeq. Quietly Mamarut stands on his toes and stares at his sick dog through the tiny window. Thinking no one is watching, he wipes away tears. The dogs howl mournfully.

Day comes. It's eight above zero and the dog is still alive. Just before leaving, Mamarut runs up on the hill and shoots his favorite dog. Later he says, "Now there will be a fight among them to see who the new leader is. Already, they are thinking of the future."

As we set off in the sleds, water splashes up on us. We hold our legs up high. The sled tips sideways, and I grab caribou skins. As I slide, Jens clamps his leg over mine and hangs on to me by the tops of my polar bear pants. The gold in the sky fades, and the air is mild. There's no good ice anywhere. Today, no one is laughing.


JULY 2004. Following the failed walrus hunt, I return to camp with the families of Jens, Gedeon, and Mamarut as they hunt for narwhal.

It's night and it's light. Instead of summer sun and quiet water, which is normal in an Arctic summer, the weather is blustery, more like northern Scotland than the polar desert of northern Greenland. Today the temperature is just below freezing. We are traveling on Jens's old red boat toward the Kangerlussuaq Fjord for a summer-long narwhal hunt. There are kayaks lashed to the deck and duffle bags with kamiks and warm clothing, despite the fact that it's July.

When the rain stops, fog drops over us like a hood. We weave blindly between moving icebergs and fight a hard southerly wind. "At this time of year the ice floes are swimming with us," Jens says. They are all melting stripes and softened craters like pelts made from mottled narwhal skins stuck to the moon.

There's a gunshot. I drop to the deck and everyone laughs. "I've been shot at a few times," I tell them with an embarrassed smile. Jens says, "It's just Gedeon telling us where to find him." He steers toward the fading sound on the far side of the fjord. We are almost at the cliff when the fog clears enough for us to see those waiting for us: Gedeon, Mamarut, another Inuit hunter named Hans Kristiensen, and their families.

The sun swings north but stays high in the sky. Mamarut and Tecummeq climb aboard and sail with us up the 50-mile-long fjord. "The narwhal come up this fjord to mate, calve, and feed," they tell me. "It's their 'cantina.' They eat the small polar cod in here. Last August some Danish biologists put a big net across the mouth of the fjord to put radio transmitters on narwhal. So far they haven't caught one," Mamarut says gleefully. "Those whales are smarter than we are, or at least, smarter than the Danes."

The narwhal was once thought to be a mythical creature. Its spiral tusk inspired the concept of the unicorn, but few had seen the animal from which the tusks came. Cups and eating utensils for royalty were made from these "horns"; they were crushed into medicinal powders thought to cure epilepsy and heart palpitations, to neutralize poisons, and to prevent fainting, rickets, and melancholy.

The "unicorn" turned out to be a small, deepwater cetacean 13 to 20 feet long that lives in frigid Arctic waters in open leads and under the ice. It is hunted from a Greenland kayak with harpoons and is the source of summer meat and mataaq for subsistence villagers in northwestern Greenland.

Jens's wooden, high-prowed boat putt-putts slowly up the fjord. We sit on deck, legs stretched out, our backs against the wheelhouse, and watch the world scroll by: black cliffs, white tidewater glaciers tumbling down to the water.

Tecummeq pours coffee from a big thermos. Squadrons of little auks fly by. A scallop of rain touches the edge of the ice cap and the skies churn gray. A storm is coming, and the first raindrop hits Jens's head. He looks up from the open wheelhouse: "This isn't our summer. We never used to have rain here. We don't know how to call it anymore."

The storm hits. Waves wash over the deck as rain comes down hard. Mamarut quickly secures the kayaks while Tecummeq stows pots and pans, cups and thermoses under the bunks below. The engine coughs, then suddenly goes dead. Wind slams the side of the boat, rocking it. I ask where the emergency gear is, just in case we get swamped. Again, the men laugh. There are no survival suits, no life rafts. We try to keep the boat headed up into the wind as Jens works on the dead engine. *"Imeq,"* he yells up to me, meaning there's water in the fuel line. He fiddles with something I can't see and eventually the engine restarts.

We chug up the fjord in pulsing rainsqualls. Mamarut squeezes in beside Jens at the wheel. They chatter away, their wet cheeks shining, ducking when the spray flies. They were boyhood friends in Moriusaq, and the charm of those early days is still evident. They don't try to get out of the weather. It's all the same to them. Jens turns, looks at me, and smiles. He gestures toward the harpoon point hanging on a rope in front of the boat's tiny windscreen. "For navigation," he says in Greenlandic, laughing. It swings back and forth in the storm like a metronome.

A piece of ice floats by. It looks like a long tusk sticking straight up. Tecummeq leans close to me and whispers: *"Qilalugaq, qilalugaq."* This is the fjord of the narwhal.

Qilalugaq, the Greenlandic word for narwhal, means "the one that points to the sky." Now we know why this name is so apt. An American researcher in dentistry, Martin Nweeia, made an amazing discovery. "The whale is intent on understanding its environment," Dr. Nweeia said. When electrodes were attached to a captured narwhal's head, Dr. Nweeia noted that "changes in salinity around the tusk produced signs of altered brain waves, giving preliminary support to the sensor hypothesis." The eight-foot-long spiral tusk—actually one of two teeth—is a sensory organ with ten million nerve endings that "tunnel from the tusk's core to the outward surface, communicating with the outside world." With it, the whale can detect "subtle changes of temperature, barometric pressure, particle gradients, and probably much else, giving the animal unique insights." And so it turns out that "the one that points to the sky" is reading the weather.

Four tents go up and the Primus is lit. Our first camp is in an abandoned Thule-era village near the top of the fjord. In earlier times, when this house site was part of a settlement, its shamans functioned as doctors, metaphysicians, and scientists, monitoring the comings and goings of life and death, weather, and the migrations and availability of animals. These turf-and-stone houses were in use here until 1999. We peer into one that, until recently, was inhabited by an old man who refused to live in a modern house. Now it is falling apart. Sitting on a drift log watching the fjord, the men tell stories: "Once when I harpooned a narwhal in Savissivik, it turned around and struck my kayak with its tusk. I had to paddle in fast before I sank," Jens says. Mamarut tells of a narwhal that stabbed his boat. "We had to stuff our anoraks in the hole to keep from sinking," he says laughing. Jens tells me that sometimes orcas come up the fjord and try to kill narwhal. "We saw an orca that had lost a fight. He was floating with a broken-off narwhal tusk sticking out of the top of his head."

Tecummeq is in charge of cooking melt-ice for tea and boiled arctic char, plus rice and soy sauce. How wonderful it is to be camping not just with the men but with whole families: Tecummeq and Gedeon's wife, Marta, and their seven-year-old boy, Rasmus, plus Marta's twin sister, her husband and children, and Masuuna, Jens's grandson by marriage.

A hunting society is a privileged way of life. "We weren't made to buy things and sell things," Jens says, "but to live altogether and bring food home to our families." Between hunts there's plenty of do-nothing time, when Gedeon can teach his son how to throw a harpoon, make a kayak, train sled dogs, skin a seal. There are no macho antics in this world. Such behavior would be against all codes of morality and conduct, and anyway, it would be too risky. Hunters learn from each other. "Our hunting tradition survives because we are still living it, and it survives through the stories we tell each other every day," Jens says.

THE STORM PASSES, and the sky becomes slate gray with an imprisoned sun trying to break through. "It has been weather like this all June, July, and now August," Mamarut says. "It is becoming winter already in August, and we haven't had summer yet." Climate change means not simply warmer weather but chaotic, intensified, unseasonable, unpredictable weather. Out on the water ice groans. The mountains are white with new snow. A river of glacier ice slides imperceptibly between two cliffs of red scree.

Hunting means waiting. Tecummeq and I lie on our sleeping bags and thumb through two dictionaries: English to Danish and Danish to Greenlandic, trying to learn new words. The children are never any trouble. They neither demand nor complain but play in the round-the-clock light, sleeping when they're tired, eating when they're hungry, playing quietly on the shore in sealskin mittens and canvas anoraks.

After dinner the hunters get busy with their gear, coiling harpoon lines and sharpening points. Mamarut inflates an *avataq*—a seal intestine that serves as a float to keep a narwhal from sinking after it's been harpooned. From now on our diurnal schedules will be reversed. We'll sleep during the day and hunt all night.

In preparation the men carry their kayaks to the edge of the water. A Greenlandic kayak is 18 feet long and sits very low in the water. The slender frame, once covered with sealskin, is now more often brightly painted canvas. The art of building this watercraft—so essential where ice travel quickly becomes water travel—was apparently forgotten by earlier Arctic peoples but was re-introduced as the last wave of Inuit hunters began migrating across Smith Sound to Greenland.

As soon as European explorers started arriving in the far north, the wood they brought from the continent became available to the Inuit people and was later scavenged from wrecked vessels. Now Greenlanders make kayaks from imported Danish wood. Ironically, the Inuit to the west in Nunavut have left the kayak behind and now use motorized skiffs and canoes.

A black cloud appears, the wind rises, and rain scatters forward like seed. The men look at the sea. "It has to be calm for the kayaks," Hans says, "because rough water is very dangerous for them. They don't swim, they don't do "Eskimo rolls," and they can easily be overturned."

We retreat to the tents and drink tea. Rain softens the hard-edged glacier ice lying on the gray stone beach. Jens and Mamarut doze. Calm water at midnight and bright skies. "This is how a Greenland summer night should be," Mamarut says on waking. The hunters pull on their kamiks and sealskin spray skirts. The fjord is ribbed with pods of seals; polar cod blacken the water near shore. The women climb to an outcrop of boulders above camp and, always vigilant, glass the water for narwhal. Far out a rainbow sticks straight up, like a tree trunk, from a cloud. Except for me, no one here has ever seen a tree. I doze in the midnight sun. A strange sound wakes me—a muffled gasping. Then Marta cries out: "Qilalugaq!"

Far down the fjord there's a slight disturbance in the water. But the women are already running down to the shore. Marta gets there first. She stands knee-deep in the water with the pointed stern of the kayak between her knees, steadying it for Gedeon as he climbs in.

Narwhal swim in pods of 20 or 30 animals. We can already hear their breathing. Gedeon pushes off and paddles hard, all sinew, stealth, and speed. He's lithe and feline, and though the narwhal are still far away, he paddles into position quickly.

An hour passes. Gedeon crouches motionless in his kayak, clinging to a piece of drift ice in the middle of the fjord. His back is to the oncoming animals, his head is bowed, his paddle is held off to one side like a broken wing. The narwhal are breathing and blowing as they come up the long fjord. They have such an acute sense of hearing that if Gedeon so much as shifts his feet in the kayak they will hear him and dive.

Now we can see them, their mottled backs humped in turbulent water. The breathing, splashing, and gasping is loud. As they approach, Gedeon bends slowly—all the way forward, his face averted so they can't see his eyes. The water is all chop and gray bodies, diving and deep breathing. A tusk lifts up as if pointing at the sun, then swords back down. Gedeon's kayak is wobbling. Water and ice splash over the foredeck, running under the harpoon. The narwhal are all around him, under him. How easily they could capsize the kayak. He waits to make his move until they have passed. Just before he begins paddling, a wind out of the north gently pushes the bow of his boat toward the whales. They see Gedeon and dive.

DAYS AND NIGHTS GO BY, but we remember only the nights. At 1:30 a.m. there is snow at the head of the fjord, a bowl where four glaciers tongue down to water, and by five o'clock every peak around us is white. "Why is bad weather chasing us?" Mamarut asks. Gedeon sits alone in front of his tent staring at the water. Seals swim by but no narwhal. Water-smoothed pieces of ice litter the beach like torsos.

Our body clocks have adjusted to what Tecummeq calls narwhal time. Bread, cheese, and jam at six in the evening, and dinner at six in the morning. After all, time in Greenland was always told by the arrivals of birds and animals, not numbered days.

Long clouds arrow up the fjord like harpoon shafts. There are no whales. Another day goes by, and another. Clouds stream by in blustery gasps. A boat comes in, anchoring in front of our camp. It's a hunter from Qaanaaq with his little girl. No one goes down to the shore to greet him. He comes into our camp bearing gifts of town food, then leaves the child with us, and motors up the fjord alone.

"He always goes by himself," Jens says, coolly. "What he just did is wrong. He's ruining the hunting by sailing up the fjord in front of us. He doesn't share. He always tries to be first, but usually he doesn't get anything. You can't be a hunter and live here alone. We don't believe in living that way. We do as our ancestors did, working and living together. If he got in trouble out in the fjord, no one would help him." Harsh words from a gentle man.

The sun makes a rare appearance at 3 a.m. Kayaks are pulled to the water and push into the silver mirror. A narwhal blows in front of camp. Gedeon and Mamarut move quickly toward it. Water flies off their paddles in wild strings of glass and gold. The water looks solid but isn't. Impermanence is the thing up here, ice becoming water and water becoming ice. Where light falls a certain way, the kayaks disappear altogether and the hunters float, half water, half men.

Five hours pass. Collapsing icebergs break. The sun lifts above serrated pink clouds. Wind ripples water like a hand crushing a page from a book. The foredeck of Mamarut's kayak turns the color of rust, as if in one day it had aged. There have been no narwhal, but time is not of the essence. "We've been drinking 20,000-year-old glacier water all our lives," Tecummeq says. "What's a few more hours?"

Nights are light blasted and so are the days.

All evening the children have been playing, catching small fish with their hands and pulling homemade toy sleds with strings up and down the shore. Just as they finally tire and go to bed, the sun completes its elliptical loop around the northern end of the sky and moves into the east until the water goes to glass and morning becomes day.

Under a warm sun, ice is breaking. Marta wakes Gedeon. She thinks she hears whales, but it's Mamarut making the noise by curling the sides of his tongue and blowing out hard. "Maybe that will make the narwhal come," he says. Gedeon pokes his head out of the tent and scans the water. Icebergs that have been pushed by wind to the far side of the fjord glisten in the sunlight. Breaking ice makes a banging sound. Mamarut puts his lips to my ear, "Qilalugaq, qilalugaq," he whispers, laughing and lunging like a narwhal. Marta gives him a look but says nothing. Again Mamarut curls his tongue and blows. This time, the narwhal come.

Gedeon is so fast he's on the water before I can find him. "He didn't even put his kamiks on," his wife tells me. He's barefoot and mittenless and the air temperature is hovering at 32 degrees.

Then I see him. Bent all the way down on the kayak's foredeck and hanging onto a piece of drift ice, he's motionless, waiting, unperturbed. An hour goes by. He hasn't moved. The narwhal are coming his way.

Turbulent water. Deep breathing. The pod passes. Gedeon bursts forward, his shoulders windmilling. He's in the middle of the whales now, the water so roiled it's hard to see the low-riding craft as the animals toss and dive. A narwhal rises up in front of Gedeon's kayak, the tusk pointing skyward. Gedeon throws the harpoon, then reaches back and tips the float into the water. *"Nauliqigaa!"* Marta cries out—he's harpooned a whale!

Red, bloodied water and a bobbing float. Mamarut appears. He's been on his way all this time. Paddling in close, he rises up, straight-backed, and throws the second harpoon.

The animal is towed toward shore. The small waves are red with blood. Gedeon stands knee-deep in the water, waiting for the incoming tide to float the narwhal in. He cuts a patch of skin and fat off the back of the whale and chews on it, then brings another piece of mataaq to his wife, scores the fat, garnishes it with soy sauce and wasabi, and shares it with the others.

The men have been sharpening their knives. Everyone helps. The head is cut off and the long tusk removed. Gedeon washes it in the fjord, and it comes out white. The narwhal was female, and she was lactating. That means her calf will probably die unless another whale adopts it. When Jens sees that I'm upset he says, "We cannot see in the water who is male and female, who has a calf and who doesn't. Now we know that's why she struggled so hard." There's a pause and he begins smiling: "Are you going to cry again?" He's not being irreverent, he's teasing me, remembering the time I cried during a polar bear hunt. "You are still like my grandson was eight years ago!"

After a dinner of rice and narwhal Jens sits pensively. Two years ago his kayak flipped over, trapping him underwater, and he almost drowned. He says he feels fear now, whenever he gets into his boat. "Sometimes I just sit in it and hang onto the side. I can't go any further." The others listen quietly. Fear and failure are treated with as much respect as bravery and hunting success.

He continues: "The elders tell us not to cross danger, to respect our fear, because it may be the thing that saves our lives. Even this morning I felt it in my chest when we went out. I get dizzy and can't tell where I am."

SUNDAY. We pile into Jens's skiff and motor over to Qeqertat, a tiny subsistence village perched on a mound of rock ten miles across the fjord. It's the only village for 50 miles. At the small inlet where we tie up, there's a stone-and-turf house, its roof studded with caribou antlers and narwhal skulls. Seal blood stains the boat ramp. Kayaks are stacked on drying racks hung with a fringe of narwhal jerky and walrus meat gone dark from the sun.

Jens takes me over the crest of a hill to see something. I look down: Hundreds of sled dogs are tethered in a small valley. Jens smiles and makes the motion for going out on the ice with them. "Soon, it will be winter again," he says, "And then you and I will be on the ice, where we are happy."

Later we join the others and stroll through the tiny village. It's here that they still remember how to make traditional string figures of Greenland's animals—a wintertime activity that has delighted Inuit children for thousands of years. Narwhal skulls crown every roof. An elegant older woman in black sealskin kamiks greets us. I'd met her the year before in the village of Siorapaluk, north of Qaanaaq. She's tiny and alert, with sparkling eyes. "I'm here in the summer and there in the winter," she tells me. Her name, Pallunuaq, means "a polar bear crawling on all fours to keep from going through thin ice."

We sit down with her at the shore in the sun. She says that at 73 her eyes are going, but when she was young she could manage anything. She was born here and says it was always a place of good hunting. "But all the people I grew up with are dead. They're all gone. Missing them makes me feel I want to go with them. It's hard."

When I ask how it was here when she was growing up, she says that the animals were fatter then and didn't have to swim as far because there was nothing to scare them, no motorboat noise.

"Our houses were made of stone and turf and heated with narwhal oil. They were warm and nice. We built the walls thick. We were never cold. Everyone had a blubber lamp burning. Ours was a foot wide, and we'd make the flame go high. If my lamp went out, I took the light from a neighbor. There was always someone taking care of the lamps."

Her father hunted every day. When he came back, she'd scrape and dry the skins, so they were ready to use for clothing. From early spring until late September they lived in tents, then they built their winter sod houses. "Our walls were lined with sealskins—not that crap we have today. Life was very nice then.

"We traveled by dogsled or kayak. My mother and I sewed the tents. That way we always had a place to live. Life was portable. Everyone had a task. Even the children. It was all skins. No cloth. There was a lot to do. I've sewed skins so much that my fingers are shaped funny. See, they are tapered. They hurt, and sometimes I wish I could cut them off. I started sewing when I was five, and I still sew all my clothes, mittens, and kamiks."

She remembers that her father was a good hunter and they always had meat. "But sometimes the tents fell over in storms. If a whiteout lasted for days, it was not easy, but it was always worth living. We had a good life."

She had never seen a shop until the small one was built here in Qeqertat. "We had only what we caught or could find. We liked best the mataaq. It is the best food I know. I'm not keen on things we get in the shops. There have been many changes in my life. That's OK. But the worst one was alcohol. It's the worst thing that was brought to our world. We could do nicely without it."

Pallunuaq met her husband when he came to Qeqertat after his forced relocation when Thule Air Base was being built in 1953. "The Americans wanted our land and they threw us out. We had nothing to say about it. My husband's mother arrived here with three small children in September when winter had already begun. No one was given a house. They made it through to summer in a tent. They were strong."

Her father told her she would make a good wife and to choose the best man around. "I was happy with the one I chose. We had eight children. The first ones, I gave birth to them alone in a tent. The rest were born with a midwife. They nursed me until I had no breasts left." She points to her flat chest. "I nursed each one of them until they were two."

Winters were spent sewing and dancing. "We had a gramophone, and we sharpened the needle like we do an ulu [woman's knife]. We went to Imina's house because it was biggest. There we danced the polka until we were dizzy; we danced the dark time away."

She says she heard many stories of shamans from her father. The white men forbade them to tell the stories, but the families told them anyway. "Once, my father was on a frozen lake looking for a snowy owl when he felt something near him. He turned and saw a polar bear down on his elbows like a human. He couldn't shoot it, and the bear didn't attack him. My father said the bear had just wanted to come near."

Pallunuaq's sons all grew up to be hunters, except one. They go north of Siorapaluk to hunt polar bears. "Things have changed a lot, but if you want to live up here, you can't manage without being a hunter. That's why I don't think hunting traditions will disappear. But I know humans are changing the world. They don't respect things as they should. When things go bad, that's the reason why. I heard they are even changing the weather."

MID-AUGUST. Ravens tumble, churning the air into a cooler season. For the first time since we arrived in July, the all-night light has begun to dim. I'm huddled on the deck of Jens's red boat coming down from the head of the long fjord. It's well below freezing. We thread our way between icebergs. I tap Jens on the shoulder and point. The sky has already begun to grow dark, and for the first time since April, we can see a new moon.

All week, Jens has been moody, worried about his dogs, which have summered on Herbert Island. "I don't think anyone has fed them this week," he says. As we round the corner and head north, we can see a red ship anchored in front of Qaanaaq's warehouses. It is the once-a-year supply ship from Denmark.

"Some years the ice was too thick and they had to turn around," Jens tells me. "Now the ice is not coming in early anymore, so we get things for the store, but we need ice more."

Town seems ridiculously large and noisy. We've been gone a long time. I notice that some people are drunk. Why? I ask Hans, who also works as one of the town's policemen. "People are drinking more than they used to here. Before, we had more murders and suicides. Now, it is not about anything. Maybe just forgetting," he says.

After a short sleep and something to eat, we head to Herbert Island. It's time to bring Jens's sled dogs home. Gedeon passes us in his kayak. He's seen orcas at the mouth of the fjord. "They scare the narwhal toward shore, so maybe I can get another one there," he says. Jens smiles. He's happy just to be seeing his dogs again. "In six weeks or so the ice will spread out between Qaanaaq and Herbert Island and up north to Siorapaluk," he says. Despite the bad ice last spring, he's still hopeful. We putt-putt toward the island. The sea is choppy and the empty dinghy towed behind his boat bounces hard.

Green meadows with hummocky grass and small ponds greet us. This is Herbert Island, also known as Qeqertarsuaq. Once it had a thriving village. Now the houses are deserted, except for one still used in summer by a Qaanaaq couple. We clamber up ice-clad rocks. Jens calls to the dogs. He's laughing now because he can see them lunging and leaping, yowling with joy. There's a calm happiness in his eyes.

Four at a time, he leads them down to the skiff. Or rather, they pull him. They know the routine and jump into the small boat. "Be good now, please don't embarrass me," he says, going back up the hill to get four more, leaving me to mind them.

When the skiff is crammed full with not an inch to spare, we head out. All 15 dogs ride happily. This isn't their first time crossing a rough channel. After a while all fall asleep but one, and he rests his chin on the gunwale, watching the water.

Though the distance appears short, the trip takes four hours. As we approach Qaanaaq, the dogs sit up and begin howling. Ilaitsuk, Jens's wife, waits at the shore. "By the time the ice begins to come back, their coats will be thick," Jens says, looking at the dogs with a calm happiness in his eyes. The kayak, the open sea, and lunging pods of narwhal are Gedeon's realm. For Jens, it's these dogs and traveling on the sea ice. A wing of spray flies over the animals' heads as we approach. They know that winter is on the way. Jens makes the gesture of snapping the whip, then moves his hands laterally, indicating the spread of ice.

When Jens whistles, the dogs look around expectantly. Ilaitsuk is waiting at water's edge. It's almost dark and the moon is out. We drop anchor and row the dinghy full of dogs ashore.

2007. It's late February and I'm in Qaanaaq. I climb the hill behind Hans and Birthe's guesthouse in the evening. Lagoons of open water are threaded with melting sea ice. I see water sky—mist unfurling like incoming waves from areas of open water and lifting into the sky. Farther south there's open water all the way to Moriusaq and Savissivik, and looking north toward Siorapaluk, open leads are like black sores.

I'm trying to take in the losses, but I can't. I'm filled with rage. Perhaps the term climate change should be changed to climate care, since it is carelessness that is bringing so many changes to life as we know it and most likely will bring much of the life of humans and megafauna on this planet to what may be the end.

The Arctic is carrying the deep wounds of the world. Wounds that aren't healing. Bands of ice and tundra that protected Inuit people for thousands of years, ensuring a continuity of language and lifeways and a meta-stable climate, have been assaulted from above and below, inside and out. Pollution, greenhouse gas emissions, the crushing demands of sovereignty and capitalism, war and religion have severed the strong embrace of ice.

Positive feedback mechanisms are piling up. Once, albedo was king. Now open water exhales mist and draws in heat that melts more ice. The Arctic traps 25 percent of Earth's $CO_2$. The equilibrium of Greenland's ice cap is skewed: ablation giving way to ablation with no more snow gain. The polar amplification effect goes against cultural survival: Poleward of 70° N, the temperature can be two or more degrees higher than the global average.

Water begets water. Absorbing solar heat, open oceans stay open; open leads in winter ice exhale mist that shelters the ice from cold, causing the open water to expand. Near the coast water undercuts the sole of a glacier's foot until it slides.

The Arctic marine ecosystem co-evolved with ice. Seasonal sea ice functions much like the Amazon canopy. It is a ceiling and a shelter; it gives nourishment and creates its own weather. Ice is a presence in the ocean. It keeps things calm and cool; it is home to 90 percent of the animals in the Arctic ecosystem.

Reductions in sea ice thickness change the mixing and upwelling of nutrients in seawater. They alter the open water ecosystem, the migration patterns of marine mammals whose corridors are being invaded by ships carrying iron ore in Nunavut; oil and natural gas in Alaska, Russia, and Greenland; radioactive waste in Russia's Barents Sea. The ecosystem's fragility is hard to see, since so much of the life occurs under the ice. But open that lid and you have ruined the lives of polar bears, walruses, seals, fish, birds, Arctic people, and whales.

The natural cycle of a tidewater glacier, even without global warming, is full of complexity. More than melting and snowfall accumulation affect a glacier's life. Having a toe in the water gives it more vulnerability. Where the glacier enters the water, it is licked by tides, driven by the shape of the coastline and water depth. The thickness and topography of the moraine can alter a glacier's stability.

So much about a glacier is true of human society. After too forceful an advance, the toe snaps and the terminus of a glacier can become grounded on its own debris.

Water is an awful duality. Water means birth and loss, it being the origin and final death of sea ice and glacier. Once ice loses its skin, texture, temperature, and shape, there's no getting it back.

*La mer de glace* shatters, and the glacier's face, its toe and snout, and the storytelling tongue.

MORNING. Gedeon comes early to tell me it's time to leave. The shore ice has firmed up, and we're going by dogsled north to Siorapaluk, the northernmost subsistence village in the world. At an average of six miles an hour by dogsled, it should take six hours to get there—an easy, straight shot up the coast made so often no one gives it much thought. But this morning, because of the warmth, the snow is wet and the ice under it is mushy.

Instead of the usual wild ride out of town, we move cautiously. The sky is almost dark, almost light. Mikele and Mamarut join us. We are lacking only Jens, because, at the last minute, he was too busy in the mayor's office to come.

We pass a half-melted iceberg stranded close to shore. The air temperature is warm, but so much open water makes the wind feel cold. There's a jag on the coast. We stop and look: The ice is all mush and water. The men talk. Gedeon and Mikele shorten the trace lines and we lunge straight up a wall of ice onto the ice foot, suspended at the edge of the coast at low tide. The sleds are heavy and the dogs struggle. A line gets caught, and as three dogs fall backward, Gedeon leaps into the middle of them up and hacks at the ice with a harpoon until the lines come free. The dogs leap; the sled slams down hard. We are on the path of ice once again.

The way is littered with frozen humps, and the ride is so brain shattering that we get off and trot behind the sleds. Trace lines tangle, dogs fall and are dragged, get up, keep running. Near the entrance to Siorapaluk, we jump down again onto the fjord ice. It's taken ten hours, a ridiculously long travel time. We arrive just as darkness descends.

A village of about 50 people, Siorapaluk sits on a hill on the northern shore of a small fjord. Five glaciers tumble down at the head. Just around the bend, on the way north, are the bird cliffs of Neqe and Pitoravik. Beyond is the historic village site of Etah, the staging ground for many North Pole attempts over the years. Siorapaluk has a small store, a tiny elementary school (grades one through eight), a chapel, and a skinning house where animal skins are prepared, tanned, and sewn.

We've rented a house where we can cook and sleep. Its windows overlook the broken shore-fast ice. "This is the second ice of the winter," Otto Simiqaq, tells me. He and his wife, Pauline, both in their 40s, are fine traditional hunters here. "In the fall we usually go north by dogsled to hunt walrus, but now we have to sail. The water is so rough, when the ice tried to come in, it came up broken by the waves. Then the ice came in and lay down flat, but after last week's storm, it broke up again."

Talk of the blizzard that almost killed Gedeon is still on people's minds here. Just before the storm hit, a hunter tied his dogs to the ice, ran up to his house to get a drink of water, and came back out to find that the fjord ice had broken up completely. He never saw his dogs again.

Eva, a Norwegian schoolteacher who has lived in Siorapaluk for 15 years, recalled it as a terrible night when no one slept. "We could not get to each other. A dog's house flew against my front door and the dog inside died. Snow filled the schoolhouse and blew into the small doghouses where females with pups find shelter. Every window was blanked out by snow. The whole of the fjord ice broke up."

After getting the snow out of the schoolhouse, Eva resumed teaching. "I teach ages 6 to 14. The population of this town varies between 62 and 50, depending on who goes down to Qaanaaq in the winter. I was hoping we could get mobile telephones, but it turns out you need 70 people to get the service. We may never have that many here. Now with the ice so bad it becomes quite isolated up here. I haven't been to Qaanaaq all year."

From the schoolhouse we walk up the hill to visit Otto and Pauline again. Pauline is sewing a pair of mittens out of sealskin bleached pale tan; the dog-hair ruff is white. "In a normal year, Otto would be up the coast hunting nanoq," she says. "But not now."

Otto comes in and pours a cup of coffee. He stirs sugar in and keeps stirring. He's tall, strongly built, and restless. "Seven years ago we could travel on safe ice all winter and get animals. We didn't worry about food then. Now it's different." He looks out at the fjord. "This is the second ice of the winter. In the fall we now have to sail to get walrus. The seas are very rough and it's dangerous. We always went to the ice edge west of Kiatak Island. Lots of walrus out there. But the ice doesn't go that far out now. The walrus are still there, but we can't get to them."

He tells me that when the moon is half full, the ocean current is safer and they can travel then. "Before, it didn't matter," he says. "Seven or eight dog teams from here would go to Kiatak together twice a month. We could get three to five walrus each time. Last year, in February 2006, I got one walrus. That was the last walrus I got from the ice."

Now they have to travel by boat, because the water is open. In the dark time it's hard to see. "On the ice, you can see everything," he says. "In winter, it is our light."

Siorapaluk usually empties out in the winter. All the men go north up the coast to Etah and Humboldt Gletscher (Glacier), whose terminus is 75 miles wide, to hunt polar bears. "There's a huge current up north that moves the ice around," Otto says. "From year to year we could travel over to Canada, to Ellesmere

Island, but that ice is gone now too. Many times I went there in the '70s and '80s. I used the drift ice to take me there. Now there's a border patrolled by Twin Otter planes. The Canadians say it is theirs, but we are all the same people."

He says that so far, the migration of little auks is still the same. They still come to Siorapaluk, Neqe, and Etah on May 10, and the local families climb straight up the cliffs with long-handled bird nets and scoop them out of the air. In the old times, birds provided food in the season between ice and open water, and they are still a good source of food. Bird-skin parkas were meticulously sewn together from the tiny skins and worn as underwear.

"We still have food, but if the weather keeps getting worse, even the future of Siorapaluk is uncertain," Otto says. I mention an attempt years before to move north up the coast. "Maybe," he says. "The weather used to be good up there, maybe it still is. But the current is stronger and the wind, and there's lots of pressure ice. Even here, where we are protected inside the fjord, the sea current is bringing big waves that are eating away our land."

Pauline has finished the mittens. She hands them to me. "Would you like them?" She asks. I look at them. Her stitching is impeccable and they fit perfectly. I nod yes and lay a stack of kroner on the table. Otto throws his cold coffee out. He's distracted and out of sorts. The usual humor isn't apparent today.

Their boys run in, grab a hunk of bread and run out again. "We have three boys, but their future as hunters is very uncertain," Pauline says. "We no longer advise them to become hunters. The climate is taking the ice, there are more quotas on animals, and the prices of things are becoming so high. So we want them to get an education."

I ask if feeding the dogs was becoming a problem, and she says that the local kommune received 24,000 Danish kroner for dog food from Denmark. "But it's only a short-term solution for a problem that will be getting worse every year."

Otto goes out. Down on the hill in front of the house he stands, looking at the ice. Pauline continues: "Many of us are behind with our debts. We are not so good in our moods now. I worry when Otto and the others go out on the ice now. It's more dangerous than it used to be. Around here, it is depression and changing moods. We are becoming just like the ice."

Evening. I climb the hill to look for my dear friend, the ever elusive Ikuo Oshima. We met in the early 1990s out on the ice. I heard about him when I first arrived in Greenland because he was one of those outsiders—he's Japanese—who immediately took to Greenlandic ways. In the first month here he mastered Kalaallisut, the Greenlandic language, and learned to drive dogs and hunt walruses and seals. That was in the 1970s. He was an accomplished hunter by the time we met again in 1996. He'd lost most of his dogs to distemper, and Jens was bringing him some puppies on our sled. He did a little jig on the ice in thanks.

Ikuo came to Siorapaluk as part of an expedition team to support the climber Naomi Uemura while he trained for his solo walk to the North Pole. Uemara came and went; Ikuo never left Greenland. He married Otto Simigaq's sister in Siorapaluk and had three children, who live nearby.

Now Ikuo lives on his own and has just moved into a new house that he and his son built. Below is the skinning house, and that's where I find him. His back is to me. I tiptoe up and put my hand on his shoulder. He turns, already smiling, and we hug hello.

Ikuo is skinning an artic fox. Everything he does is artful and quick. His forearms are thickly muscled and smooth. He sits on a block of wood positioned on the seat of a chair so he doesn't have to stand at the table to work. He's lashing the fox's tail to a stick with a reindeer thong. The radio is on and Sting is singing. Two young girls peer in shyly. "That's Matthew Henson's great-great-granddaughter," Ikuo says, nodding at one of them. Matthew Henson was the African-American first mate who accompanied Robert Peary to the North Pole and, like Ikuo, quickly learned Greenlandic ways.

After Sting, Ikuo spins toward me on his block of wood. "My son and I are thinking of moving someplace maybe in future years. We might consider moving north. It is terrible when the ice goes out and it's too dark to hunt by boat. We are trained as ice hunters, and without ice it's terrible. If global warming was coming gradually, maybe there would be some chance to adapt, but it's so sudden, this coming of climate change, we are in a panic.

"So I have to think, how can we survive? The game animals on the land are OK, so maybe we will have to be land hunters. But then we'll need vegetables, because only marine mammals have all the vitamins and minerals we need. But vegetables don't grow here. We've always gotten what we needed from the seals and whales.

"Now fish are coming north. We found salmon in our seal nets. It is all very surprising. There are insects now—mosquitoes and black flies—and we never had them before. We have sickness in the animals, and back home in Japan bird diseases have started.

"Maybe I'm coming old enough, and I'm thinking, we have enough work and food and we must be thankful for the day. When the ice is bad, I go trapping. I walk all the way up these canyons where the glaciers used to be. There's a snowy owl living up there. Many beautiful things to see. But some of the young people are thinking only of tomorrow, that they can go get money somewhere, but it never happens. It takes many years and still you have no money," he says laughing. He pulls the fox skin inside out, then hangs it up to dry.

276

"I was maybe 25 or 26 when I first moved here. First time was 1972. Then left, but came again in 1974 and stayed forever. My nationality is Japanese because my mother is still alive, but she has never visited. I was studying engineering at a university in Tokyo, but I found it to be too organized—no freedom inside of it. See, it's already fixed up for you, and you just have to choose some numbers to make the thing go together right. That's not for me!

"At the same time I found I liked nature more than mechanics. I climbed many mountains, and I still wanted to have some feeling of freedom. We can decide for ourselves how to live. . . . But maybe I want too much. . . . I don't know."

He invites me to have coffee. "Look at this. It's my new house. Maybe it helps me somewhat in my old age!" Laughter. In the basement are hundreds of skins tanned and hanging—arctic fox, arctic hare, and a musk ox skin.

We climb back up to the porch and look out. It's dark again, but the snow is bright. "There were polar bear tracks right there. It walked past the village in the night. We don't want to stay outside much at night anymore because there are so many bears. They've come onto the land because the ice is so broken, and it's dangerous. Maybe my house will protect me!" Laughter again.

"More and more there are reindeer and musk ox. If there were a thousand last year, there are 4,000 this year. We hunt them because there is not enough food from the sea. On June 10, I'll be 60. My mother will be 93. She's still doing something every day and grows lots of flowers in the springtime. So I hope I will be that way too. Ten years ago I visited Japan. Oh, it was so terrible! All the houses and buildings had changed. I couldn't find my house. I came into a kind of panic and thought, I'm a wild man. What am I doing here?

"Here, we own no land. We only make an application to put a house on it. Maybe it's better this way, because we use it for only a short time! My wife is ill, and she lives in Qaanaaq now. But my granddaughter was living with me for a year. She just left. This is the first month without her, and I miss her. Between us there was no need for words."

We drink our coffee in silence and look out at the ice. I invite him to our house for dinner, and he accepts eagerly. Someone brings reindeer meat from a neighbor, and we make Scandinavian-style stew with dried fruit. I ask Ikuo again about the changing climate and the ice. He says, "I heard in Savissivik there have been many visits by polar bears. The ice condition is not good, and there are many bears in the village. The males and females without cubs are traveling about all winter. The season for polar bear hunting is from December to June, but only Greenland's best hunters can hunt them. The ice is so bad only one bear has been killed in the entire northern district.

"When we go to Humboldt Glacier to hunt nanoq, we need two days to cross the inland ice because the sea ice in front of the big glacier is still not good. Up there is a small island that's closer to Greenland than to Canada, but the Canadian government wants to own it. They are planning to build an air force base. They want more power to protect the Northwest Passage, because the ice is melting all the way to Alaska. The world is peaceful on its own, if only we didn't bother it. People don't need power. When I look at our dogs running, that is their happiness, and for all of us, to eat enough, that is happiness as well.

"Every year we go over the inland ice to Etah. We used to be able to travel up the coast a ways, then go over at Pitoravik. Now that is not possible. We have to go to the end of our own fjord and up onto the ice that way. That's how bad the sea ice has become. It

is becoming more and more difficult to travel. We used to get walrus up there. Now we get musk ox. But the meat of both animals
is good. Oh, it's so beautiful up there! Yes, I think we will survive
somehow, maybe just on beauty."


BACK IN QAANAAQ I brood and walk. The poet Joseph Brodsky
said that the purpose of evolution was beauty. Up top, the melting ice cap gleams. And rough ice between town and Herbert
Island looks like Hiroshige-style frozen waves. Up on a hill I find
Birthe. Her arms are crossed. "I'm so cold," she says, as if she were
dying. No one here ever complains about the cold. Heat—*kiak*—
yes, but not cold. She tells me why: With so much open water,
the mist that rises from it adds humidity, making the air feel cold.
Now 40 or 50 below begins to seem unbearable, whereas in the
"dry days" no one was bothered. "The soles of the hunters' kamiks
are always wet now," she tells me.

Together we gaze out at the ruined ice. Ruined or not, it is still
beautiful. "We love it here. We never tire of watching the icebergs
and the light. It changes all the time. It was always calm here. We
always had calm minds too. That's how we faced the weather. But
now the weather is not right. Now the ice isn't behaving."

Years ago, a group of people tried moving farther north to
live in the old way without anything from the stores, without
schools or medicine, but it failed, Jens told me earlier. When I
asked why, he said, "They ran out of coffee and sugar!" I walk the
narrow paths to the kommune office to see him. Walls of hard-
packed snow lie against the west walls of the houses—remnants
of the big blizzard. At Jens's house Mamarut and Tecummeq are
visiting, plus Ilaitsuk's daughter, who has just had her fifth child.
"It seems to be all she wants to do," Ilaitsuk says. She's a bold,

broad-shouldered woman who doesn't mince words. The daughter bears children; the grandparents—Jens and Ilaitsuk, who are only in their 40s—take care of them.

Jens dandles the baby on his knee, so very gently. He's a big man, and next to him the baby seems impossibly small. Until recently, no one but Hans at the Qaanaaq Hotel had a television. Now Jens and some others have them. Television in Greenland is perhaps a model of what it might be elsewhere, limited programming with a blank screen until evening. At night Danish news is followed by news from Greenland's capital city, Nuuk, then there's a movie from Europe or the United States, dubbed in Danish with subtitles in Greenlandic. "Before we had televisions, we had storytelling," Jens says. "Now we just tell our ancestor stories out on the ice."

When the television goes off, there's talk in the room of the economic plight of hunters who cannot get enough food, who cannot afford to buy a boat. Last year Greenland's prime minister, Hans Enoksen, had come to talk to the hunters in Qaanaaq. He is from a village too, but farther south. The townspeople told him they had no jobs, and the hunting was not bringing enough food to live. Every Thursday they get 300 Danish kroner from the kommune office. "And three hours later, they've spent it," Ilaitsuk says. "The prime minister listened, but after he left we never heard from him again."

Since the early 1990s the grocery store has quadrupled in size. Television, long-distance telephones, faxes, and Internet service became available for those who could afford it. And more recently a bar. "We didn't need that. We didn't need alcohol at all," Ilaitsuk says.

It has been decided to call a meeting of the hunters in town, so they can report on the effects of the declining ice. Hans Jensen generously allows us to use the guesthouse at the hotel. I want to beg forgiveness from my Greenland friends for the vandalism and greed of the so-called developed nations, where only profit counts, where decisions are not made with the biological health of the planet in mind but only the material wealth of a few. Where true poverty is enforced on even the wealthy, where the social standard is to live and work in a fixed place. Where the moral implications and social injustices of the crushing demands of extreme capitalism are daily overlooked and denied. Where the "I" is king; where the "getting" is done at the expense of others; where the concept of "we" is considered a form of weakness.

It was here, in Greenland, that I had my first taste of true civilization: where the demands of capitalism are held to a minimum. Where the conscious choices of what an ice age society might keep of its traditions were added to what they deemed useful from the 21st century—harpoons and cell phones; helicopters and kayaks; and far beyond a simple striving for survival, gratitude for the natural beauty of the place and its importance in their daily lives.

Here, people's basic needs are met first; matters of buying and selling come last, if at all. This is a society based on sharing, self-discipline, patience, modesty, resilience, flexibility, and humor, and a sophisticated understanding of the circular continuity, the transmission of ecological knowledge that keeps tradition alive.

Sunday night, meeting time: Jens and Ilaitsuk; Mamarut and Tecummeq; Gedeon and Marta and their son, Rasmus; Paulus, a hunter I hadn't met before; the elder Qav; Otto S.; and Toku Ishima (Ikuo's very capable daughter, in her 20s). The faces are solemn. There is none of the usual joking and laughter.

"I'm so sorry for what is happening to your ice," I begin. "As you know, it is happening not because of anything you've done, but because of what our countries have. Many of us are working hard to slow down carbon dioxide emissions, but we are not as powerful as we would like to be. I apologize for this. I'm so sad. You have been so kind to me and many others like me who have come as visitors here. I will work hard to bring your situation to the ears and eyes of other people, so they will know the precious culture that is being lost. I have tried to show them who you are—there's no one quite like you in the whole world."

An embarrassed silence, then slowly the hunters begin to talk. Otto begins:

"Even in summer it's hard to get out to the islands or up the coast by boat because the sea is so rough. The whole village of Siorapaluk used to empty out in February and March. We all went north to hunt polar bears. All the way to Humboldt Glacier and Washington Land. The ice was good. Now we have to go up and over the ice cap to Kap Alexander, and it is very strenuous and a danger to our lives."

Mamarut: "And going south is very bad now too. We used to go over Politiken Glacier or around by the coast. Now, even these ways are gone. Many years in July we could get to Kiatak on the ice. I remember the dogs' paws used to bleed where they went through the meltwater into the ice."

Tecummeq: "I always went with Mamarut. The ice was always solid. It wasn't something we thought about. It was always very cold. Town is like prison. I used to be a hunter's wife. I worked with Mamarut 100 percent of the time until a few years ago. I wanted to help him be a hunter. We had fewer animals to get because the ice edge was stronger. Now we can't even reach Kiatak anymore, and that's where the hunting was best. If the ice was still good, we could still be taking care of ourselves. There are only two of us. It must be very hard when there are a lot of children."

Otto: "Until 2003 the living of a hunter family was possible. But not after. In 2004, that's when the ice got really bad. The wives have to get jobs, so they are able to buy fuel for a boat. That's the only way for us to get walrus now. We share the boats, but still it is more than we have. We did not want these things. They didn't interest us. Now because of the ice condition, we are forced to go this other way."

Toku: "The situation is like this: To be a full-time hunter is impossible today. In the old times the whole family worked together with no outside work necessary. Now even full-time hunters like Otto, Mamarut, Gedeon, and the others have to take care of dogs and family, and it's hard because of the changes. The hunters' wives have to get jobs to pay rent or bills. There are always those new kinds of expenses now. The hunters' sons are told to get an education because it's the only way to make a living in the future."

Jens: "Farther south the village of Savissivik is now very bad. They can't even go to Kaviok because the ice is so thin. They had polar bears in the fjord between Kaviok and Savissivik, but they couldn't get to them. Before it was the place with the most solid ice. We used to go there when the ice was gone here. We'd always go south. When the ship came in late summer, there was still ice and they had to used dogsleds to bring the supplies in. Still, there are seals to hunt down there, so as for now, they have food for people and dogs. But they can't go very far in any direction except by boat, and at Moriusaq, they can no longer get out to Saunders Island and the other small islands where they hunted walrus. And if there's no ice, there's no walrus."

Paulus: "Summer weather has changed. It used to be so quiet. Lots of sun. Now it's stormy. We have rain and wind. We've never seen rain before. We didn't know about it. Up the big fjord you can sometimes hunt seal, but there's only water in front of Qeqer-tarsuaq. The glacier is gone at the head of the fjord. That's why our sea level is rising."

Jens: "The warming is causing those glaciers to calve huge icebergs, but then they melt. That didn't used to happen."

Toku: "It's not just the current but also the Gulf Stream coming north from the coast of Africa. I've been following the changes on the Internet. The Gulf Stream is dividing. It is now pointing at Herbert Island. Yes, the warm part of the current is coming this way, and when it comes this way, it brings different fish, but it takes the ice away."

Jens: "When you think of having to make money, something we didn't have to think about before, the main resource is seal—year-round. But even for seals the conditions are worse. When living in open water, the hides are not as good. There are pollution problems too, and the seals sink when you shoot them because they are skinny. Now the ice period is so small, the seals' new coats don't change. They don't molt. Even in winter they don't change their coats. Now the furs aren't good for making our clothes."

Toku: "We have pollution problems, too. POPs [persistent organic pollutants] attach to the fatty tissues of all the animals, the seals, walrus, and polar bears. In other areas the POPs go up in the air, but here it's too cold, so they sink and invade the fat that we need to eat to keep warm."

Otto: "The ice melts so fast now. It snows on the ice more often and makes the ice melt. Now we notice something new about the animals. They have yellow blubber—it should be whitish pink—and the livers look unhealthy. Maybe it's because they are not clean anymore. If we didn't have a store and halibut in the fjord, we wouldn't have enough food. We'd be getting hungry now."

Jens: "Only last year we received information about the health of people and animals in east Greenland. They have the highest levels of mercury contamination in the world.

"We are also told that Greenland is melting faster than expected. Both the inland ice and the sea ice. The next 10 to 20 years are critical. Even if the world makes changes in emissions now, it will take 100 years for the ice to come back. If we wait 20 years to make these changes, it may never come back. But by then, who will remember how we have lived these many years?"

BLOWING WIND and the village is quiet. The waxing moon is three-quarters full, and steam from the houses drifts with falling snow. New ice has formed. It bends with the tides and changing currents. "In a few days the moon will be full and the weather will be stronger," Mamarut says. "Maybe wind waves will break up the ice. Maybe it will stay."

Up the hill the edge of the ice cap is dull white. Once it shone like a gigantic diamond. It stood for the icy fastness and cultural continuum that is the polar north and seemed indestructible. Now scientists worry about the accelerated erosion of the "big ice," as the meltwater in a glacier's natural drainpipe, its moulin, drills down to bedrock and lubricates the base, causing great hunks of ice to slide.

People here say there is life in every part of the Earth, animate and inanimate, and maybe moving ice is one aspect of that dynamism. Only 70 years ago the coast of Greenland was thought to be inhabited by beach spirits called Ingnerssuit, who looked like humans but had no nose and tried to lure people to live with them. Giants called Sarqiserasait rose out of the rough seas. Traveling in a half kayak, they killed anyone in umiat or kayaks.

Female giants had long claws on hands and feet so strong they could dig holes in granite. They lived in the solitary fastness of the coastal mountains. The Isserqat lived in the ground and winked sideways. They were known for tickling people to death. The Tarrajarssuit were invisible spirits. If their shadows fell on you, you would instantly die.

A humped cloud rises. Sun squeezes between two rumps of mist. The Earth's veins now run with poisons, not only radiation but the more widespread industrial and agricultural organic compounds that ride the winds around the world and drop down in rain and snow onto the massive snowfields and ice sheets of the Arctic.

POPs travel vast distances, accumulating and condensing in snow, ice, and water. Insoluble and long-lived, these toxins pass through the food web quickly: from seal, walrus, beluga whale, and narwhal to polar bear, bird, caribou, musk ox, and human through breast milk and meat. Sixteen percent of the north Greenlandic population have high levels of toxins in their blood. There is transboundary pollution in all eight circumpolar nations.

A polar bear eating a ringed seal passes along the poisons of an industrial world to her tiny cubs; Greenlandic women who breast-feed their children have the highest concentrations in the world of methyl mercury in their breast milk. POPs endanger the immune system and provoke a susceptibility to diabetes and all kinds of cancers and impair hormonal and reproductive functions, which is why hermaphroditic polar bears have been found in Svalbard and Greenland and Inuit males have abnormally low sperm counts.

In earlier times one of the local shamans might have intervened. "Shamans were our scientists. Like the narwhal, they knew the environment, they could read the weather," a hunter

said. Now scientists are reading the climate and saying that surface reflectivity is at a premium because snow and ice turn 80 percent of the sun's heat back into space, whereas the dark surface of
open water absorbs 90 percent of the sunlight. "Even if there were
shamans in Greenland," a Danish scientist said, "climate models
show that carbon dioxide emissions at almost 400 parts per million and getting higher are overriding Sila. And they are predicted
to rise very soon to 450 ppm and higher." It's not just an environmental injustice but a social and spiritual one as well.

Later I walk the shoreline with Mamarut. A friend we pass
along the way says the ice is getting good again, that he was able
to go out at Herbert Island and saw narwhal but the ice was too
broken to hunt. Narwhal in February? I ask. Are they usually this
far north? "They think it's April," Mamarut says.

A gold bullion sun spills over fractured ice and open water.
There are legends about Arctic winters when the ice never comes,
but never one about endless summer. Strands of ice stretch out
into golden threads. "What can we sew with that?" I ask "Money,"
Mamarut replies, grinning slyly, It's good to see him make a joke
again. But he's serious: "We never needed money before. Now we
need boats with motors to hunt when there is no ice. But we have
no money to buy them."

It begins snowing again. Snowing for no reason. It covers rotting ice, warming it, causing it to melt further. Jens joins us. The
wind is noisy. We pass by the big diesel-generating plant that gives
lights to the town.

I'm wearing the sealskin kamiks made for me by Jens's wife,
Ilaitsuk. Two older women stop to admire them. Extra sealskins
were once sold by the hunters to Denmark. Jens says: "When the
natural living took place, the prices for these skins was high. In the
1970s when I started as a full-time hunter, there was no problem.

The little money we needed could be made directly from selling the skins. We kill and eat about 30,000 seals a year. That's more skins than we can use. So we sold the surplus to Denmark.

"Then Brigitte Bardot and Greenpeace started talking about seals. She thinks we are the ones clubbing baby seals, but we are not. We would never do that. We are not commercial seal hunters. People like her don't know who we are. She is still harming us today. We are no longer allowed to sell sealskin outside the country. She has made us poor, she has disrupted our families. We used to live all together out on the ice. Our wives, who always traveled with us, making clothes, preparing skins, working at camp, helping with the dogs, and teaching the youngsters, now have to stay in town and work at the few low-paying jobs available."

We cross the bridge to Jens's house, where, with the onset of increased precipitation, even in a polar desert, there have been flash floods. He looks at his dogs, pointing to the ones that have pulled our sled on various trips. He's smiling, but when he turns to me, he looks lost, as if wondering what has happened to his world. It's too much to take in, to understand, to accept, I want to say, but lack the words in Greenlandic.

Later Jens tells me about a cave south of Pituffik where the polar bear spirit is still "very strong." He says he would like to take me there sometime. In the old days he might have been a shaman instead of a mayor. I ask him again about the time he was called by the polar bear spirit and he says, "I still fear it; I would still turn away from the call because we live in modern times. There is no place in society for a shaman now."

Ilaitsuk gives Jens a look. She knows everything that he is and could have been. She feeds one of her many grandchildren a stringy piece of polar bear meat. Mamarut and Tecummeq come by. Under his parka, Mamarut is wearing a T-shirt with a picture of a baby on the front, though he, Jens, Mikele, Hans, and others who grew up in Moriusaq or Dundas are unable to have children, and no scientist, doctor, or shaman has been able to make things right.

In January 1968 a U.S. B-52 bomber carrying four nuclear weapons caught fire and crashed on the sea ice in Bylot Bay, eight miles west of the Americans' Thule Air Base. The nuclear payload—each bomb was 1.1 megatons—ruptured on impact, and radioactive materials including uranium, plutonium, and americium from those weapons were dispersed across the ice and into the sea. The Americans had been storing nuclear weapons at their base and for use in their "Chrome Dome," Cold War exercises—all against a Danish law that prohibited such weapons in its territory.

It took 700 men nine months to clean up the site. Only three of the four bombs were actually accounted for. The fact that one weapon was missing was only recently revealed. Many on the cleanup crew, including Inuit hunters and Danes, were not provided with protective clothing, nor were they monitored for signs of radiation sickness subsequently.

The contaminated ice and the wreckage were removed and shipped to the U.S. After that, subsistence marine mammal hunting north and south of the base resumed. The exposure to contaminated meltwater and the consumption of marine mammals that live in the region "could be" the cause of radiation illnesses, sterility, and cancer, a Danish doctor said.

I'VE MADE A SIDE TRIP south to Nuuk to visit my friend Aleqa Hammond. I haven't seen her since our long walrus-hunting trip in 2004. I've known her for 17 years. Now she is married to a

Danish geologist and has an adopted son, a hungry boy they found camping on their doorstep, and she's Greenland's minister for finance and foreign affairs.

Aleqa is tall, with a villager's sturdy frame and a beautiful face. She is notoriously strong minded, with a mixed Inuit/English heritage, and straddles the worlds of subsistence hunters, Danish royalty, and Greenlandic politicians with ease. Her family in Uummannaq was impoverished, but it was there that she learned to handle a team of dogs and make decisions on her own, which has served her well in handling powerful politicians.

"I never keep back from anyone. I always go from the heart," she tells me. "I got it all from my mother and grandmother. My father went through the ice when I was seven. My grandmother was an old woman then, a wise woman. She always helped people, told them out loud how to make their lives better."

Aleqa was 12 when she was left to care for her two younger brothers after her mother went off to work in the Black Angel mine. "I only saw her a few times a year, so I had to take charge. We were poor—she is still poor. If my father had lived, I'd probably be less outspoken!"

In cabinet meetings in Greenland and Copenhagen, she is brutally frank. She'll look a member of the Danish or Greenlandic Parliament in the eye and say, "You didn't mean that, did you?" "That's how I handle them when they're trying to maneuver things their way." Her opinion is listened to. She knows what it's like to be from a hunter family; she knows what is happening to the ice.

"The most important thing we can do is to make the transition between old and new without losing the old ways," Aleqa says. "We can't avoid capitalism. It is coming at us from all directions. But what we take of the new world and how we match it to the best of the old, keeping the spirit intact, is very important. The way we have been living together is the essence of a living tradition."

South of the Arctic Circle and the hunting life, Nuuk is a sophisticated Arctic town with a museum, a university, two newspapers, a large hospital, a publishing company, a bookstore, restaurants, and discos. The night of the lunar eclipse we're invited to a minister's house for a party. As the moon slips into its black envelope, someone says, "It is like what is happening to our ice," then goes inside to eat Thai food and sip French wine. These aren't stuffy affairs but rather vivacious and young at heart. Aleqa is spearheading a push for Greenland's independence. "We have always wished for this, and now the time has come," she says.

Unlike other Arctic nations, Greenland has a majority population that is Inuit. Colonized in 1721, the island was a Danish protectorate until 1978. Now Greenland has home rule, with a thoroughly Inuit Parliament, but trade and foreign affairs are still overseen by the Danes. Greenland's economic stability is based on the infusion of 680 million Danish kroner per year by the Danish government. When I asked a politician in Copenhagen why they are so generous, he said, "Guilt. Greenland is our national treasure."

Petrol, heating oil, lumber, building materials, and housing are heavily subsidized, making it possible to live on a hunter's income. Medicine and education are free. What will Greenland do without these generous subsidies? What will be the consequences of this claim for independence for an island of rock and ice and a tiny population of only 57,000 souls?

As the white lid of ice comes off the top of the world, elements of temperate ecosystems will move north. Vegetables and crops of hay have been grown in south Greenland below the Arctic Circle since the Vikings landed there. Ocean fishery is thriving. Warm-water cod are repopulating Davis Strait, and pockets of lead and zinc have been uncovered near the town of Uummannaq. Gold and diamonds have been found, and oil companies predict that the northern waters may hold at least 31 billion barrels of oil and gas.

"We are now on the path to independence, and nothing can stop us," Aleqa says defiantly, perhaps unaware of the devastating toll the extractive industries take wherever they go. "With all this mineral wealth, we won't need the subsidy. There will be nothing that ties us. We will be economically free." When I tell her that such freedom is a myth, she ignores me.

QAANAAQ AGAIN. The sky flames out in sun dogs, frost glitter, and zodiacal light. The wild stirring of the ice cap that is everywhere in the news is nowhere to be seen. Yet if it melts completely, sea level will rise 23 feet around the globe.

Climate is always changing, and ice is on the move. When the North American plate carried Greenland toward the Equator, Greenland was warm and lushly vegetated, and the temperature of the Arctic Ocean peaked at 73°F. Once, there were dinosaurs at the Poles.

Sea ice is ephemeral and so is the ice sheet, but on another order of geologic time. Some 125,000 years ago, in the last interglacial, the ice cap was smaller than it is today. The cyclical restraints were still in place then: An interglacial always gave way to an ice age, and ice ages were more numerous than warm times. But the sun's cycle was such that it gave off less heat than it does now, and with anthropogenic climate forcing, the stable interglacial time we've been enjoying is now crumbling, with no upswing into a new ice age—where we should be headed—in sight.

Not "warming," but deadly heat. The ice sheet is melting, breaking, surging, acting more like an ocean than a mountain of ice, as if mimicking what it is about to be.

The paradise called the Holocene is ending, and a new epoch, tentatively named the Anthropocene, is beginning—an era when climate will be forced against its cyclical "instinct" to become cold again. A warming world has all kinds of consequences. As the continental ice sheet retreats, botanical explosions will occur. Populations of terrestrial northern animals like musk ox and caribou will continue to increase, and the ice-free land will rebound and rise like bread, shaking off all vestiges of "a place called winter" and the men and women who thrived there.

But on this cold February day at almost 78° N the ice between Qaanaaq and Herbert Island has come in and is holding. For a moment I remember how Greenland used to be—a place wrapped in stillness but for echoing, howling dogs. And the vast, unpolished blank of ice giving way to I don't know what—an unfurling of some sort. Emptiness filling with ice.

Standing on the hill behind town, I tip my head back. The ice cap stirs memory: a memory with no images. Only a vast, unpolished blank, gessoed by sun, a canvas waiting for paint, for stories.

"But there are no stories up there," Jens tells me when he sees me glassing the ice sheet's gleaming edge. A Danish scientist once told me about the life on the ice cap, though it's microscopic. "It's a dusty world," he said. He was referring to the particles of cosmic dust from outer space that drift down on the ice cap. Because dust is dark, each particle absorbs heat and drills tiny holes in the ice, which in turn capture more dust. Some cryoconite holes are as deep as eight inches.

AFTERNOON. Sun travels fast in snowy gauze, looking more like a moon. Moon is nowhere to be seen. The fluted palisades of rock and ice are horizontal ribs under snow-rich peaks. The blue spires of retreating glaciers are knocked down as they defy gravity, ascending the mountain instead of dropping down its flanks. Inside the ice cap, moulins drain meltwater in secret until the sole of the glacier's foot begins to move. Ice walking until it is water.

Up the coast I can see where the braids of a frozen river are coming undone, its silt-loads fanning into seawater. The stretch-marked glaciers are falling. Sea ice is being pulled apart in ribbed stress fractures, the bones of ice that can no longer hold the bones of men. Ice flattens the torment of the sea. Without it, the sea tosses like a tree in the wind.

Whether I'm in Greenland or not, I dream about it: Gedeon is walking around carrying huge panes of ice with a simple curved handle. He seems perplexed. He walks out into the water and lays pane after pane on the tormented sea, but the ice-windows keep breaking. Again and again he tries, but open water keeps lapping at his knees, and the more panes he lays down, the more water appears all around him.

A hunter has killed his own dogs during the night because he didn't have the food to feed them. They lie in their harnesses, motionless. The snow that touches them has melted back an inch or so, and the bodies are outlined by pale blue ice. Beyond, the dogs belonging to Jens, Mamarut, Mikele, and Gedeon are staked out on rumpled shore-fast ice. There should be 4,000 or 5,000 dogs out here, but there are only a few hundred, their nightly chorus of howls gone almost silent.

The sea ice covering of the Arctic Ocean has declined by an area equal to Texas and Arizona combined, and the remaining ice has lost six feet of its thickness. Soon the Arctic will be open water all summer, and the elusive North Pole will be a floating nub, an arbitrary mark on a map made by the humans who have demolished its ice hold.

Hans Jensen takes me around town to say goodbye to friends: Ilaitsuk, Tecummeq, Gedeon, Marta, Rasmus, but I can't find Mamarut or Jens. I look up as we walk and see a narrow cloud hanging over a shard of ice. The two move in unison as if married, then the cloud slides away, and under it, the ice shard melts. Toward the familiar islands, Qeqertarsuaq and Kiatak, there is now a world of imeq, open water, with no winter white for the moon to shine on. Just heat sinking down, reaching up to grab at ice from beneath.

I remember hurtling across translucent sea ice with Jens, the dogsled fishtailing as we gained speed. We were laughing and clinging to each other to keep from falling off. Now single icebergs tilt in turquoise moats, reminding me that the wobble and tilt of the Earth—the Milankovitch cycle—should be propelling us toward an ice age. Some winters have been colder the last few years, perhaps because when there has been almost no sunspot activity, the weather cools off. Plus there has been a persistent La Niña that keeps Pacific waters cool. But most winters have been erratic and mostly too warm, with the average global temperature still rising.

Toward Kiatak sea smoke curlicues into a pale sky. "The sea water is boiling," Hans says. "That's why there's smoke. The water is warmer than the sky. The world is on fire. How can ice ever come back to a place like this?" he asks.

Then we do find Jens on the road to the airport. He's been looking for me while we were looking for him. We stop in the middle of the ice road to hug goodbye. I ask how he's feeling about things since the meeting. He looks at me and says, "I no longer want to live to be an old man."

After checking my bags, we go back to town, since it will be another three hours before boarding time. At the shore we find Mamarut, preparing to go hunting for musk oxen. He coils his green lines and lays four harpoons under the lash ropes. He'll travel over the ice cap to a region south of Savissivik. It is a dangerous trip to take alone, but he has to find food. As sea ice diminishes and glaciers retreat, the narrow valleys along the coast are growing more vegetation each summer. But by the time young caribou are old enough to reach these distant valleys, the grass has already reached maturity and is declining in protein. As a result, the calves do poorly.

Mamarut stands by his sled. "We are marine mammal hunters, but now we are hunting on land for whatever we can find—caribou, musk ox, birds," he says. He looks out at the ruined ice, at the long bands of open water glittering in the sun, his eyelashes white with rime ice. Putting his hands to his head like horns, he makes a sound that's more bovine than marine mammal. The brown mark of frostbite on the side of his face has blossomed like a dying flower. He stashes a rusty rifle and an extra pair of kamiks under the lash rope and pushes the sealskin headband up on his head—a black crown for the tragic-comic Mamarut. His wild eye looks skyward. Snow comes down like chipped light. He grabs his harpoon, jabs it into the ice, and grandly holds it at an angle like a staff.

"When we have nothing," he says with a strong lilt to his voice, "we flourish. That's how it's always been for us here."

# ACKNOWLEDGMENTS

>>> <<<

HEARTFELT THANKS to the people of the circumpolar north whose voices and thoughts gave this book life in a difficult time of abrupt climate change. In Alaska, special thanks to Joseph Senungetuk, Catherine Senungetuk, Herbert Anungatuk, Winton Weyapuk, and those who lent their voices and thoughts in Wales, Shishmaref, and Nome.

In northern Russia, thanks to Andrei Volkov, biologist, guide, and enthusiastic translator, and to my Komi friends.

In Nunavut, thanks to John MacDonald, Carolyn MacDonald, Leah Otak, Zach Kunuk, Sonia Gunderson, and Mitch Taylor.

In Greenland, deep and heartfelt thanks to my friends in Qaanaaq, especially Jens Danielsen, Mamarut Kristiansen, Gedeon Kristiansen, Mikele Kristiansen, and their families, and Hans and Birthe Jensen. In Siorapaluk, thanks to Otto and Pauline Simigaq, and as always, to Ikuo Oshima. And thanks to Aleqa Hammond.

This book was made possible by a generous grant from the National Geographic Expeditions Council. Special thanks to Rebecca Martin, who makes all things possible, and to Lisa Thomas of National Geographic Books.

Thanks also to Karen Merrill and Williams College Environmental Studies, Clemma Dawsen, Brendan Kelly, and Rita and Jaimie, Pat and Mark, Thekla and Callum, and John McGough for all kinds of help and moral support.

This book is for Tom.

# NATIONAL GEOGRAPHIC
# SOCIETY EXPEDITIONS COUNCIL

≫≫≫ ≪≪≪

NATIONAL GEOGRAPHIC'S Expeditions Council, which supported Gretel Ehrlich's fieldwork for this book, is a grant program that was launched in 1998 to fund projects involving exploration of largely unrecorded or little-known areas of Earth, as well as regions undergoing significant environmental or cultural change. Ehrlich's investigation into the effects of climate change on peoples of the Arctic is an outstanding example of the kind of work we fund, and of a timely story that needs to be told.

Indeed, the story is central to all of the expeditions and fieldwork that the Council supports. The remarkable results of these often far-flung efforts enlighten and inspire our global audience through the coverage produced by our broad range of media—from magazines and television programs to books, Web, lectures, exhibits, and educational products.

While the Society has a 120-year history of supporting scientific fieldwork—with more than 9,000 grants awarded, largely through its Committee for Research and Exploration—new grant programs have emerged in recent years to support a variety of projects that also fulfill our mission of inspiring people to care about the planet. These programs include the Expeditions Council and Conservation Trust, Young Explorers and NGS/Waitt grants, the Legacy and Afghan Girls' funds, and the All Roads Film seed grants, not to mention our Education Foundation grants. (Visit *nationalgeographic.com* for more information regarding these programs.)

Landmark among the hundreds of projects funded by the Expeditions Council is conservationist J. Michael Fay's 15-month, 2,000-mile "Megatransect" through the central African rain forest in 1999-2000. The magazine stories—including extraordinary photographs of a deeply threatened wilderness—and other media produced inspired then Gabonese president Omar Bongo to set aside 11,000 square miles of his country's land area as 13 national parks, which are still intact and under continued development. This is the most positive outcome we could hope to realize from our support and coverage.

More recent projects include Jim Balog's Extreme Ice Survey, which generated time-lapse photography of glacial melt as hard evidence of global warming; Kira Salak's adventurous journeys through Libya, Iran, and Bhutan; Tim Samaras's first ever measurements and images inside tornados; Dan Buettner's search for and investigation of the longest-living peoples around the planet; Dan Fisher's studies of a remarkably well preserved baby woolly mammoth; and amazing observations of blue whale behavior led by John Calambokidis. Also worthy of mention are the deep-sea explorations of Robert Ballard, as well as major finds by paleontologist Paul Sereno and high-altitude archaeologist Johan Reinhard. From the frigid journeys of renowned polar explorer Børge Ousland and Sylvia Earle's Sustainable Seas Expeditions to Mike

Fay's recent Redwood Transect and the uncovering of pre-Buddhist murals and texts in the cliff caves of Upper Mustang, Nepal, by Broughton Coburn and team, one realizes that the possibilities for discovery still abound, even close to home.

A young Jane Goodall received her first grant from National Geographic in 1961 for her work with chimpanzees. At the time, Jane was unable to secure funding from any other source for her unprecedented studies. She went on to receive the largest number of grants of any individual supported by NGS, and of course became an icon in the world of conservation. Every year we continue to fund hundreds of individuals, and they continue to break new ground in the fields of research and exploration—sometimes beyond our wildest imaginings. And in fulfilling our mission, we will continue to share their myriad inspiring discoveries with you.

*Rebecca Martin*
*Director, Expeditions Council*

# BIBLIOGRAPHY

## GENERAL CIRCUMPOLAR

### Books

Fagan, Brian. *The Great Journey.* Thames and Hudson, 1987.

———. *The Great Warming.* Bloomsbury Press, 2008.

———. *The Little Ice Age.* Basic Books, 2000.

———. *The Long Summer.* Basic Books, 2004.

Flannery, Tim. *The Weather Makers.* Atlantic Monthly Press, 2006.

Grim, John. *Indigenous Traditions and Ecology.* Harvard University Press, 2001.

McCarthy, Allen P. *Indigenous Ways to the Present.* Canadian Circumpolar Institute Press, 2003.

———. *In Search of Nature.* Island Press, 1996.

McGhee, Robert. *Ancient People of the Arctic.* UBC Press, 2001.

———. *The Last Imaginary Place.* Oxford University Press, 2005.

Meyewski, Paul, and Frank White. *The Ice Chronicles.* University Press of New England, 2002.

Norman, Howard. *Northern Tales.* Pantheon Books, 1980.

Pielou, E. E. *A Naturalist's Guide to the Arctic.* University of Chicago Press, 1994.

Vaughan, Richard. *The Arctic: A History.* Allan Sutton Publishing, 1994.

Wilson, E. O. *The Future of Life.* Vintage Books, 2002.

Young, Steven B. *To the Arctic.* John Wiley & Sons, 1989.

## Journals, Research Stations, and Websites

### Bellona Foundation
*bellona.no/bellona.org*
A multidisciplinary international environmental nongovernmental organization based in Oslo, Norway.

### British Antarctic Survey
*www.antarctica.ac.uk/*
A component of the Natural Environment Research Council. Based in Cambridge, United Kingdom, it has undertaken the majority of Britain's scientific research on and around the Antarctic continent. It now shares that continent with scientists from more than 30 countries.

### Canadian Ice Service
*ice-glaces.ec.gc.ca/*
A branch of the Meteorological Service of Canada (MSC) and a leading authority for information about ice in Canada's navigable waters

### Dansk Polar Center
*www.dpc.dk/*
Knowledge and service center for scientists and institutions that deal with polar research and arctic matters and for the public.

### Dot Earth
*dotearth.blogs.nytimes.com/*
A *New York Times* blog that reports on natural resources, the environment, climate change, and sustainability.

### EarthWire UK
*www.earthwire.org/uk/*
A daily overview of the environment in the United Kingdom as reported in the media.

**Fridtjof Nansen Institute**
*www.fni.no/*
An independent foundation engaged in research on international environmental, energy, and resource management politics.

**NASA Earth Observatory**
*earthobservatory.nasa.gov/*
Current information about climate and the environment.

**NASA Goddard Institute for Space Studies**
*www.giss.nasa.gov/*
A laboratory of the Earth Sciences Division of NASA's Goddard Space Flight Center and a unit of the Columbia University Earth Institute. Research at GISS emphasizes a broad study of global climate change.

**National Oceanic and Atmospheric Administration**
*www.noaa.gov*
A federal agency focused on the condition of the oceans and the atmosphere.

**National Science Foundation**
*www.nsf.gov/*
An independent U.S. government agency responsible for promoting science and engineering through research programs and education projects.

**National Snow and Ice Data Center**
*nsidc.org/*
Part of the Cooperative Institute for Research in Environmental Sciences at the University of Colorado at Boulder. NSIDC supports research into our world's frozen realms: the snow, ice, glacier, frozen ground, and climate interactions that make up Earth's cryosphere.

**Nature (subscription req.)**
*www.nature.com/nature/*
The international weekly journal of science.

**New Scientist**
*www.newscientist.com/*
International science magazine covering recent developments in science and technology.

**Norwegian Polar Institute**
*npiweb.npolar.no/english*
Norway's central institution for research, environmental monitoring and mapping of the polar regions.

**Potsdam Institute for Climate Impact Research**
*www.pik-potsdam.de/*
Research institute working on questions of climate change, climate impact and sustainable development.

**RealClimate**
*www.realclimate.org/*
A commentary site on climate science by working climate scientists for the interested public and journalists.

**Science Daily**
*www.sciencedaily.com/*
Breaking news about the latest scientific discoveries.

**Tyndall Centre for Climate Change Research**
*www.tyndall.ac.uk/*
Organization that brings together scientists, economists, engineers, and social scientists to develop sustainable responses to climate change through transdisciplinary research and dialogue.

**University of Alaska, Fairbanks**
*www.uaf.edu/*
Alaska's top teaching and research university and home of the Alaska Center for Climate Assessment & Policy.

**University of East Anglia-Climate Research Institute**
*www.cru.uea.ac.uk/*
Research institution concerned with the study of natural and anthropogenic climate change.

## INTRODUCTION

Broecker, Wallace S., and Robert Kunzig. *Fixing Climate.* Hill and Wang, 2008.
Lovelock, James. *The Ages of Gaia.* Bantam Books, 1988.

————. *The Vanishing Face of Gaia.* Basic Books, 2009.

Margulis, Lynn. *Symbiotic Planet.* Basic Books, 1998.

Morton, Oliver. *Eating the Sun.* Harper Collins, 2007.

Ward, Peter. *Under a Green Sky.* Smithsonian Books, 2007.

## ALASKA

Burch, Ernest S., Jr. *The Inupiaq Eskimo Nations of Northwest Alaska.* Fairbanks: University of Alaska Press, 1998.

Ellanna, Pikonganna, Muktoyuk, Omiak, Kasgnoc, Pullock, Sirloak. *King Island Tales.* Alaskan Native Language Center / University of Alaska Press, 1988.

Fienup-Riordan, Ann. *Boundaries and Passages.* Norman: University of Oklahoma Press, 1994.

————. *The Living Tradition of* Yup'ik *Masks.* University of Washington Press, 1997.

Fitzhugh, William W., and Susan A. Kaplan. *Inua, Spirit World of the Bering Sea Eskimo.* Smithsonian Institution Press, 1982.

Johnson, Charlie. *Nanoq: Cultural Significance and Traditional Knowledge Among Alaska Natives.* Alaska Nanuuq Commission, 2005.

King Island Native Community. *King Island Tales.* Alaska Native Language Center, 1988.

Krupnik, Igor, and Lars Krutak. *Our Words Put to Paper.* Arctic Studies Center / Smithsonian Institution Press, 2002.

Krupnik, Igor, Rachael Mason, and Tonia W. Horton. *Northern Ethnographic Landscapes.* Arctic Studies Center / Smithsonian Institution Press, 2004.

MacLean, Edna Ahgeak. *Inupiallu Tannillu Uqalunisa Ilanich: Abridged Inupiaq and English Dictionary.* Alaska Native Language Center, 1980.

McCartney, Allen P., ed. *Indigenous Ways to the Present: Native Whaling in the Western Arctic.* Edmonton, Canada: Canadian Circumpolar Institute Press; Salt Lake City: University of Utah Press, 2003.

Nelson, Edward William. *The Eskimo About Bering Strait.* Smithsonian Institution Press, 1983.

Oozeva, Conrad, Chester Noongwook, George Noongwook, Christina Alowa, and Igor Krupnik, eds. *Watching Ice and Weather Our Way.* Arctic Studies Center / Smithsonian Institution Press, 2004.

Rasmussen, Knud. *The Alaskan Eskimos.* Copenhagen, Denmark: Gyldendalske Boghandel, Nordisk Forlag, 1952.

Ray, Dorothy Jean. *The Eskimos of Bering Strait 1650-1898.* Seattle: University of Washington Press, 1975.

Senungetuk, Joseph E. *Give or Take a Century.* San Francisco: Indian Historian Press, 1971.

Senungetuk, Vivian, and Paul Tiulana. *A Place for Winter.* Anchorage: Ciri Foundation, 1987.

Smith, Kathleen Lopp, and Verbeck Smith. *Ice Window: Letters From a Bering Strait Village 1892-1902.* University of Alaska Press, 2001.

Spencer, Robert F. *The North Alaskan Eskimo.* Smithsonian Institution, Bureau of Ethnology Bulletin 171, 1959.

Vanstone, James W. *Point Hope.* University of Washington Press, 1962.

Wohlforth, Charles. *The Whale and the Supercomputer.* North Point Press, 2004.

# NORTHWESTERN RUSSIA AND SIBERIA

## Books

Anderson, David G. *Identity and Ecology in Arctic Siberia.* Oxford University Press, 2000.

Bobrick, Benson. *East of the Sun.* Poseidon Press, 1992.

Bogoras, Waldemar. *The Chukchee.* Johnson Reprint Company, 1909.

———. *The Eskimo of Siberia.* AMS Press, 1975.

Brodsky, Joseph. *Less Than One.* Farrar, Straus, and Giroux, 1986.

———. *On Grief and Reason.* Farrar, Straus, and Giroux, 1995

Forsyth, James. *A History of the Peoples of Siberia.* Cambridge University Press, 1992.

Golovnev, Andrei V., and Gail Osherenko. *Siberian Survival.* Cornell University Press, 1999.

Gorbatcheva, Valentina, and Marina Federova. *The Peoples of the Great North*. New York: Parkstone Press, 2000.

Kendall, Laurel, and Igor Krupnik. *Constructing Cultures Then and Now*. Smithsonian Institution, 2007.

Kertulla, Anna. *Antler on the Sea*. Cornell University Press, 2000.

Vitebsky, Piers. *The Reindeer People*. Houghton Mifflin, 2005.

———. *The Shaman*. Little, Brown, 1995.

Young, Oran, and Gail Osherenko. *Polar Politics*. Cornell University Press, 1993.

## Films

Golovnev, Andrei. *Pegtymel*. Yekaterinburg, Russia: Ethnographic Bureau, 2000.

———. *Way to the Sacred Place*. Hanover, New Hampshire: Dartmouth College Institute of Arctic Studies, 1997.

# NUNAVUT

## Books

Balikci, Asen. *The Netsilik Eskimo*. Waveland Press, 1970.

Beckett, Samuel. *The Unnamable*. Grove Press, 1955.

Bennett, John, and Susan Rowley. *Ugalurait*. McGill-Queens University Press, 2004.

Breton, Pierre. *The Arctic Grail*. Lyons Press, 2000.

Brody, Hugh. *The Living Arctic*. University of Washington Press, 1980.

Georgia. *Georgia, an Arctic Diary*. Hurtig Publishers, 1982.

Hall, Charles Francis. *Life With the Esquimaux*. Tuttle, 1970.

Jenness, Diamond. *People of the Twilight*. University of Chicago Press, 1959.

Krupnik, Igor. *Northern Ethnographic Landscapes*. Arctic Studies Center / Smithsonian Institution, 2004.

Krupnik, Igor, and Dyanna Jolly. *The Earth Is Faster Now*. Arctic Research Consortium and Arctic Studies Center / Smithsonian Institution Press, 2002.

Lyon, Captain G. F. *The Private Journal of Captain G. F. Lyon 1821-1823*. Barre, Massachusetts: Imprint Society, 1970.

MacDonald, John. *The Arctic Sky*. Royal Ontario Museum / Nunavut Research Institute, 1998.

McGrath, Melanie. *The Long Exile*. Harper Perennial, 2007.

Rasmussen, Knud. *Across Arctic America*. G. P. Putnam's Sons, 1927.

———. *Intellectual Culture of the Iglulik Eskimos*. Copenhagen, Denmark: Gyldendalske Boghandel, Nordisk Forlag, 1929.

———. *Report of the Fifth Thule Expedition 1921-1924*. Copenhagen, Denmark: Gyldendalske Boghandel, Nordisk Forlag, 1929.

Wachowich, Nancy. *Saqiyuq*. Montreal, Canada: McGill-Queen's University Press, 1999.

## Other Media

**Isuma Aboriginal Television**
*www.isuma.tv/*
An independent interactive network of Inuit and Indigenous multimedia.

**Nunatsiaq News**
*www.nunatsiaq.com/*
Nunavut's territorial newspaper in the Nunavut and Nunavik regions of the eastern Canadian Arctic.

# GREENLAND

Born, Erik W., and Jens Bocher. *The Ecology of Greenland*. Nuuk, Greenland: Aage V. Jensens Fonde, Ministry of Environment and Natural Resources, 2001.

Christensen, N. O., and Hans Ebbesen. *Thule*. Arktisk Institut, No. 2, 1985

Fortesque, Michael. *Introduction to the Language of Qaanaaq, Thule*. Institut for Eskimologi, 1991.

Freuchen, Peter. *I Sailed With Rasmussen*. New York: Julian Messner, 1958.

Gad, Finn. *History of Greenland*. Vol. 1. London: C. Hurst and Co. Publishers, 1970.

# BIBLIOGRAPHY

Gad, Finn, and Gordon C. Bowden. *History of Greenland.* Vol. 2. London: C. Hurst and Co. Publishers, 1973.

Gronnow, Bjarne. *Paleo-Eskimo Cultures of Greenland.* Danish Polar Center, No. 1, 1996.

Harper, Kenn. *Give Me My Father's Body.* Steerforth Press, 2000.

Heide-Jorgensen, Mads Oeter, and Kristin Laidre. *Greenland's Winter Whales.* Nuuk, Greenland: Ilinniusiorfik, 2000.

Kane, Elisha Kent. *Arctic Explorations.* R.W. Bliss, 1869.

Macmillan, Donald. *Etah and Beyond.* Houghton Mifflin, 1927.

Malaurie, Jean. *The Last Kings of Thule.* University of Chicago Press, 1985.

———. *Ultima Thule.* W.W. Norton, 2003.

Rasmussen, Knud. *People of the Polar North.* New York: J.B. Lippincott, 1908.

———. *Posthumous Notes on the Life and Doings of the East Greenlanders in the Olden Times.* AMS Press, 1976.

# INDEX

>>>— —<<<

313

# INDEX